GEOGRAPHERS
Biobibliographical
Studies

VOLUME 31

GEOGRAPHERS: BIOBIBLIOGRAPHICAL STUDIES

GEOGRAPHERS
Biobibliographical Studies

VOLUME 31

Edited by Hayden Lorimer
and Charles W. J. Withers

on behalf of the
Commission on the History of Geography
of the International Geographical Union and the
International Union of the History and Philosophy of Science

B L O O M S B U R Y
LONDON • NEW DELHI • NEW YORK • SYDNEY

Bloomsbury Academic
An imprint of Bloomsbury Publishing Plc

50 Bedford Square
London
WC1B 3DP
UK

175 Fifth Avenue
New York
NY 10010
USA

www.bloomsbury.com

First published 2012

British Library Cataloguing-in-Publication Data
A catalogue record for this book is available from the British Library.

ISBN: HB: 978-1-4411-8624-9

Typeset by Newgen Imaging Systems Pvt Ltd, Chennai, India
Printed and bound in Great Britain

Contents

The Contributors

Hugh Clout is Professor Emeritus in the Department of Geography at University College London and a Fellow of the British Academy.

Suzanne Daveau is Professor Emeritus in the Department of Geography at the University of Porto and Researcher at the Centre for Geographical Studies in the University of Lisbon.

Avril Maddrell is Senior Lecturer in Geography at the University of the West of England.

Eric Pawson is Professor of Geography in the Department of Geography at the University of Canterbury, Christchurch, New Zealand.

Introduction

The nine essays which make up this thirty-first volume of *Geographers: Biobibliographical Studies* examine the geographical lives and works of a Briton, a Portuguese national, six Frenchmen and, although he was born in Yorkshire in England, a New Zealander by adoption and residence. As our essayists reveal, however, nationality provides no easy and certainly no simple causal category either to understand or to explain these geographers' lives and works: to read them in that way ignores the international dimensions to their work. Neither does simple categorization into 'human' or 'physical' geographer do our subjects justice. Such labelling does not readily speak to that study of the connections between the physical world and its human inhabitants which was the aim for many. Nor does it speak to the fact that, for some, a career begun with interests in one theme changed over time, or to the fact that in helping establish departments and teaching programmes as several did, specialist research in one area or another of the emergent subject was complemented (if not hindered) by the generalist demands of teaching across the range of the discipline. Five of our subjects were born in the late nineteenth century, four in the early twentieth. Collectively, they shared the experience of involvement in the formalization of geographical curricula and in the institutionalization of their subject in respective universities and countries: they shared it, and, as our essayists severally show, they also profoundly shaped it. In here prefacing the detailed accounts which follow, we trace the principal features of our subjects' geographies and lives before turning to comment on three significant connecting narrative threads: recognition of geography's contribution as a science of 'civics'; the experience of war; and that sense of transnational commitment to geography across national and thematic boundaries evident in our subjects' participation in the affairs of the International Geographical Union (IGU).

The first female professor of geography in the United Kingdom and a distinguished scholar of early modern geography, Eva Taylor has before now long been a notable gap in the coverage provided by *Geographers*. Taylor was a pioneering figure in the history of geography and geographical knowledge, in effect bringing together that-then-also-nascent field the history of science with study of geography's mathematical practitioners and practical navigators, principally for periods before the eighteenth century: (although her 1966 *The Mathematical Practitioners of Hanoverian England, 1714–1840* extends into the nineteenth century, her focus there as in her earlier monographs was on mathematical and philosophical instruments and on their makers and users, hardly at all with scientific instrumentation or with the emergence of science as regulated epistemological procedures in the 'modern' sense of the term). This was genuinely innovative scholarship for which Taylor was recognized in her lifetime (despite some contemporaries' reservations about her lax standards of bibliographic citation – ironically, a charge she was fond of levelling at others). Her attention to geographical knowledge in practice and her almost prosopographical survey of instrument makers was world leading. The history of geography was then in its infancy. Historical geography was likewise, and in Britain, the United States and in France, was much taken up with the reconstitution of synchronic landscapes and explanation of diachronic change. But she was not alone a pioneering historian of geography. Perhaps because she had discerned the place of geography in practice in the past, and certainly because she was a combative proponent of geography's role in education and in identifying future possibilities vis-à-vis spatial patterns and optimal distributions, Taylor was also involved in planning new post-war geographies in England after 1945.

The inclusion of the essay on Orlando Ribeiro is likewise welcome, in pointing to the significance of Ribeiro's work and in adding to the presence (albeit one that is still under-represented) of Lusophone geographers within *GBS*. In the period covered by Eva Taylor's early modern geography, the navigators of Portugal and Spain had an almost global pre-eminence. By Ribeiro's day, Portugal's global geographical reach had diminished, burdened by her colonies' demands for independence and, latterly, by their expectation of continued assistance in a postcolonial context even as Portugal's economy and political life was restricted under Salazar's dictatorship. Ribeiro's studies of Portugal's national and international dimensions must be read against these shifting geo-political circumstances. For Ribeiro, Portugal's geography was not marginal-European but Atlantic-global, the country a node for connections between Europe, Africa and South America.

The first of the French geographers here reviewed, Aimé Vincent Perpillou, produced his major thesis in geomorphology. But Perpillou's most significant contribution was his work in regional and rural geography, notably on the Limousin region of France's Massif Central. In this, as the examples of Antoine Vacher, René Musset, Jean Dresch and Daniel Faucher each illustrate, Perpillou's work was part of an intellectual 'model of practice'. In French geography, this was rooted in an understanding of physical geography, usually at a regional or a local scale (*pays* and *paysage*), upon whose 'base' the human landscape was revealed, 'laid open' so to say, according to the sources available to reconstruct patterns of human occupation and systems of land management for a variety of historical moments while usually addressing the present day. If, and in summary outline, this is and was Vidalian geography, we must beware seeing such geography as necessarily national either in focus or result. If it was anything, such geography was innately chorological: in its undertaking, it required skilled observation, time in the field and empathy towards people, place and practice. Its results usually afforded an integrative synthesis of a region's character, from geology to *genres de vie*, and in its teaching, promoted a holistic vision for geography in both teacher and pupil and in the French geographical imagination. This is not to say that Vidalian geography was conceived as a national project: the picture of France that emerged derived from close-grained regional studies of the type produced by Perpillou, Vacher, Musset and the rest. And our study now of what we might think of then as 'Vidalianisme', the pursuit and practice of Vidal's geographical vision, needs to recognize the strength and depth of those personal influences and institutional connections which shaped the French geographical world in the first 30 or so years of the twentieth century. As is revealed here, men followed one another into teaching posts, dedicated their theses to influential teachers as they themselves influenced others and so became dedicatees, each sustaining a singular intellectual vision with regional variants while so doing. Eva Taylor and Orlando Ribeiro both saw these standards – or De Martonne's anyway – as things to be aspired to, international 'bench marks' of intellectual standard. René Musset may have been the Last Vidalian as he is here described (and the longest-lived), but Vidalianisme lingered longer.

For Jean Dresch, a scholarly life begun with physical geography in high places ended in high geographical office. His exposure to arid geomorphology in the mountains of the High Atlas was also a first-hand encounter with human poverty and the miseries consequent upon French colonial rule. Foreign field work was a formative influence in his becoming a communist. And because Dresch was a communist, the American organising committee denied him access to the International Geographical Congress (IGC) in Washington in 1952. But it was Dresch's geographical standing and not his political beliefs that determined his election to the presidency of the International Geographical Union 20 years later. Vidalian, Frenchman, citizen, geomorphologist, communist, committed internationalist, geographer? For Dresch, and for his five French counterparts, simple labels do not work.

Kenneth Cumberland's life-long sense of geography as the study of regions and his commitment to geography as a discipline of consequence stemmed from an enduring affiliation to Hartshornian chorology and to his exposure to the 'new world' of New Zealand and its geography. If New Zealand in general and its South Island in particular represented a 'geographical laboratory', Cumberland soon enough became aware of the fragility of the landscape on and in which he was experimenting. Like Dresch, study of physical geography led to recognition of the practical consequences of process-based studies: specifically, consideration of the role of soil erosion, its importance as a 'natural' phenomenon being unwittingly accelerated by unthinking human land use. It also led to Cumberland's lengthy and parallel engagements as a widely respected broadcaster, and practitioner of both local body politics and farming.

There is, then, a common bond in the ways many of our subjects helped establish geography in new institutional settings, and there is also a strong and shared sense of disciplinary zeal, a passion for instruction in classroom and in the field and a belief in geography's analytic worth and in its redemptive possibilities. This came, perhaps, from the very newness of the situation in which they found themselves and the concomitant freedom to develop their teaching programmes. It came, too, from belief in what we might think of as the civic and exchange value inherent in a geographical education and in the importance of bringing geographical thinking to bear on public issues. This is evident in their teaching: virtually all our subjects here wrote the teaching texts and the mapping manuals through which they instilled literacy, graphicacy and the benefits of comparative explanation in their students. And it is apparent in the ways in which their geography had civic consequence. Taylor addressed questions of urban planning and land management. Dresch and Cumberland cautioned about the deleterious consequences of inappropriate land management: for native herders in Morocco's mountains and of settlers' over-exploitation of high country and hill grazing lands in New Zealand. And a common theme addressed by Ribeiro for Portugal and by our French subjects was what they saw as the 'emptying' of the countryside with the result not just of changing patterns of land use and vegetation cover but the slow collapse of rural civic culture. If, and arguably, contemporary geographers pay scant heed to the impact of their work in the public realm, there are lessons to be learned from the lives scrutinized here.

What is also clear from these essays is how much the lived experience of doing geography and committed views on the value of a geographical education and of geographical thinking was forged by war. Teaching during the Blitz as her male colleagues were called into active service, Taylor contributed to the planning of post-war Britain. Vacher, Cholley and Faucher produced geographical reports as part of the French war effort in World War I. Later, Faucher used his university's facilities in Toulouse to offer safe haven for refugees from Spain's Civil War. Dresch, a combatant in World War II, took René Musset's position in Caen from 1942–6 following the latter's deportation by the Nazis following accusations of spying: as Hugh Clout notes, Musset evidenced formidable powers of recovery upon returning to find his town, university and research papers destroyed. Such circumstances may explain our protagonists' views of geography as a means to help build a better post-war world. In such biographical fragments lie rich possibilities for further reflection upon geography's role as an active subject in times of war: less, perhaps, to do with geographers' contributions to map work during conflict and to international treaties at war's end, but more prosaically if no less meaningfully, of geographers coping with war, with individual's ideas and lives changed by war. Cumberland's career, for example, was very clearly shaped by New Zealanders' responses to World War II and their desire for improved geographical knowledge and training.

Did experience of war produce generations (of geographers) committed to internationalist civic perspectives, determined never again to let national supremacies dictate others' views of a shared humanity? Such is possible, of course, but we can only speculate here. What is certain is that many of our subjects considered the meetings of the International Geographical Congress an important platform for collaborative geographical exchange. Cholley was present at all the IGC meetings in the 1930s: Paris (1931), Warsaw (1935) and Amsterdam (1938). The Lisbon IGC of 1949 was important to Ribeiro, Perpillou and to Dresch. Dresch, who was also present at the IGC meetings in the 1930s, was President of the IGU for four years from 1972. Cumberland was on the Union's Executive Committee for eight years in the 1960s. Did our subjects meet and exchange views at the IGC? How were national practices of geography understood by geographers of other nations and scholarly traditions? Here, too, we need to know more about the role played by the IGU and the IGC meetings, as sites and moments of geographical exchange and as an institution through whose functioning geography crossed borders.

Hayden Lorimer
University of Glasgow

Charles W. J. Withers
University of Edinburgh

Eva Germaine Rimington Taylor

1879–1966

Hugh Clout and
Avril Maddrell

After graduating in chemistry, Eva Taylor taught science and then studied geography at the University of Oxford, where she was trained in the holistic conception of the discipline, embracing physical, human and regional study, field observation and map work. She then wrote successful textbooks for schoolchildren, while holding part-time lectureships in London. In 1921, she joined the staff of Birkbeck College, in the University of London, where she would spend the rest of her career. In 1930, she was appointed to the chair of geography in that college and, as such, was the first female professor of geography in the British Isles. She pioneered research into the history of geographical thought and practice, submitted important evidence on the location of industrial activity, which would be of great value in national planning after World War II, and then, in a very active retirement, turned her energies to research in the history of science and the history of navigation. Her final monograph was published shortly before her death. Possessing a remarkable intellect and acute sense of scholarly rigour, Eva Taylor could be a formidable critic and a challenging opponent in the small interwar community of British geographers, inspiring both admiration and trepidation among its members.

Education, Life and Work

Eva Taylor was born at Southwood Lane, Highgate, north London, on 22 June 1879, daughter of Charles Richard Taylor, a solicitor with an office in the City, and Emily Jane (née Nelson). Three years later, Emily left her husband, who would prove to be a strict father to Eva and her older brother and sister. Toys and pets were forbidden at home, but Eva developed an early interest in garden plants and nature in general, and immersed herself in her studies. Initially she was tutored at home before going to Camden School for Girls and then to the North London Collegiate School for Ladies (then located in Camden Town), where she was an exemplary pupil, excelling not only in her work but also at singing and playing the piano (Anon. 1968, 182). Both schools had been founded by Frances Mary Buss (1827–94), the educational pioneer, who was at the forefront of campaigns for establishing schools for girls, and for girls to be allowed to sit public examinations and to study in universities. She was founding president of the Association of Head Mistresses (1874–94), and was involved in establishing the Teachers' Guild in 1883 and the Cambridge Training College for Teachers in 1885. Miss Buss was a strong advocate of women's suffrage. At the conclusion of secondary schooling, Eva won a scholarship to study at Royal Holloway College for Women and achieved a University of London first-class honours degree in chemistry in 1903.

At this time, it was usual for most female graduates to enter the teaching profession. Eva Taylor was no exception. For two years, she taught science at Burton-on-Trent School for Girls where the clarity of her well-prepared and interesting lessons made a very positive impression on the pupils and on the headmistress. In 1905, Eva moved to the Convent School in Oxford, where she combined her teaching with further study at the University in order to enhance her qualifications. Initially, she intended to work for the Diploma in Education. Miss Cooper, a tutor to women students at Oxford, advised her to consider geography rather than science as a career path, since the infant discipline was seeking well-qualified teachers (Maddrell 2009, 171). In 1906, Eva enrolled at the School of Geography in the University of Oxford to work for the Certificate in Regional Geography, and in 1907 she proceeded to study for the Diploma in Geography (Freeman 1980a, 97). Her lecturers were Andrew John Herbertson (1865–1915), H. O. Beckit (1874–1931) and Nora MacMunn (1874–1967), Halford Mackinder (1861–1947) having resigned the Readership in Geography in 1903 in order to serve as director of the London School of Economics (on Herbertson, see *Geographers* Vol. 3; on Mackinder, *Geographers* Vol. 9). Eva passed both sets of examinations with distinction and in 1908 began to work as a research assistant for Herbertson who, as Beckit would later record, had regarded her as 'the most able and brilliant of many women students trained under him at Oxford' (E. G. R. Taylor Papers, British Library MSS 69467 543A, Letter from Kate Ruddy, 3 October 1905; Maddrell 2009, 170).

Two years later she returned to live in London, which coincided with her relationship with Herbert ('Bertie') Edward Dunhill (1884–1950), second son of Alfred Dunhill who had recently started what proved to be a financially successful company selling tobacco, pipes and other products for smokers. In 1905, Bertie had married V. Penley at Wandsworth, but within a few years he turned his affection to Eva with whom he shared a house, 34 Oakley Street, in fashionable Chelsea. In 1912, Eva's first son, Spencer, was born and was recognized by the Dunhill family. A second son died in infancy, and in 1919 at the age of 39, Eva gave birth to a third son, Anthony (known as 'Tan'), who was given the Dunhill name, although Eila Campbell reports doubts about his paternity. Eva wrote of her feminist principles to her school contemporary Dr Marie Stopes: 'I have taken (from conviction!) the rather drastic step of having children without being legally married to their father, so that I can sign myself, Yours Sincerely, Eva G. R. Taylor' (Marie Stopes Papers, British Library MSS Add 58738, f. 24, E. G. R. Taylor to Marie Stopes, 23 March no year given). Eva's strict father would not recognize her sons and income from his estate would be stopped after her death. It seems that Taylor had to adopt her own sons in order to become their guardian (Anon. 1968, 182; E. G. R. Taylor Papers, British Library, MSS Add 58738, f. 24, Eila Campbell to Dr. D. Webber, 20 September 1981). Eva Taylor and Bertie Dunhill seem to have drifted apart. In 1920, Eva bought the lease of the house in Chelsea from Dunhill, suggesting that either their relationship had come to an end, or that she wished and had the means to be financially independent. The lease would expire during World War II. For the last quarter century of his life, Bertie suffered ill-health resulting from the effects of tuberculosis. He went to live in Italy, and died in Milan in 1950. Through his will, trust funds were left to the Dunhill Medical Trust, established in 1986 to provide monies for the furtherance of medical knowledge and research.

Following the death of her infant second son, Eva taught part-time for two years (1916–18) at the Clapham Training College for Girls (on the North Side of Clapham Common) and at the Froebel Institute (in Colet Gardens, west London). In 1920, when her third child, Anthony, was only in his second year, she was appointed to a part-time lectureship to teach human and regional geography at the East London College (forerunner of Queen Mary College, University of London), where she worked alongside the geologist H. G. Smith, who had responsibility for physical geography (Sheppard 1994, 4). In 1921 Eva Taylor became a part-time lecturer at Birkbeck College (formerly the Mechanics Institute) located since 1885 in Breams Buildings, off Fetter Lane, in the City of London. Leonard Brooks (later Inspector of Schools for the London County Council) lectured part-time on historical geography in 1919, and the names of William Henry Barker (1882–1929) and Gundred Helen Lydia Savory (born 1881) are mentioned as lecturers in geography in various Minutes of the Board of Studies and in issues of *The Calendar for Birkbeck College*. Barker moved to the Readership in geography at the University of Manchester in 1922; 7 years later, he died at the age of 47 (Anon. 1929, 225). The geologist Leonard Frank Spath (1882–1957) was listed among the Birkbeck geographers from 1921 to 1928; E. J. Orford was demonstrator in geography from 1923 to 1928 (Spath 1982).

Birkbeck College offered evening and weekend classes to teachers, City workers and others who wished to obtain a degree but were employed during the day. Geography had been taught there by George Goudie Chisholm (1850–1930) (from 1895 to 1908) (on Chisholm, see *Geographers* Vol. 12). Lionel Lyde (1863–1947), professor of geography at University College London, delivered lectures at Birkbeck College during the subsequent session (1908–9), and in 1908, John Frederick Unstead (1876–1965), was appointed as a part-time lecturer (Wise 1975, 1) (on Lyde, see *Geographers* Vol. 30). Since 1905, Unstead's main employment had been at Goldsmiths College in south-east London, which specialized in training teachers. Only a year older than Eva, Unstead had also acquired his geographical credentials at Oxford, and in 1912 had submitted work to the University of London on climatic limitations on wheat cultivation in North America for the degree of DSc (Econ), although his main interest remained in geographical education (Gilbert 1966, 335). His post at Birkbeck College became permanent in 1914 (Anon. 1966, 199). Five years later, Birkbeck became a constituent college of the federal University of London, and Unstead was appointed to its newly created chair of geography in 1921. He was 'diligent and industrious but lacked the lively mind of his collaborator and colleague, Eva Taylor' (Campbell 1987, 48).

Taylor taught many aspects of the discipline at Birkbeck College. Between 1922 and 1924, for example, she taught the history of geographical discovery; historical geography; and world geography: special regions. From 1929–31, she also lectured on the physical basis of geography; geomorphology; climatology and oceanography; and map work. From the early 1920s she served as an examiner on several papers taken by geography students enrolled in the various colleges of the University (Archives of the University of London, AC 8/24/1/1: Minutes of the Board of Studies in Geography). She realized too that for her career to progress, she needed to develop a research profile to complement her teaching and examining skills. She never fully revealed the reasons behind her choice to specialize on discovery and exploration and the development of geographical knowledge, but according to her pupil Eila Campbell (1915–94):

> She believed that scholarly writing depended on the appraisal of original source material. It is possible that she was pointed to her study of geographical thought by A. P. Newton (Rhodes Professor of Imperial History in the University of London)

> who was himself interested in the geographical thought of the Middle Ages and who she would have come to know as a fellow member of the University's Board of Studies in Geography. She may also have been attracted to her subject by her early training in the natural sciences. (Campbell 1987, 48)

A wealth of manuscript sources was available at the Public Record Office (then located in Chancery Lane, just a few paces from Breams Buildings), and the Reading Room and the Map Room of the British Museum provided an abundance of material within easy walking distance of Birkbeck College. Teaching for 3 hours from 6 o'clock most evenings was complemented by intensive research activity during the day as well as managing her household and family.

As was the custom of that time, Eva submitted published work and further work in progress to obtain a DSc in 1930 under the title 'Studies in Tudor Geography'. Her dossier must have included the typescript (perhaps even the proofs) of her forthcoming *Tudor Geography, 1485–1583* to be published by Methuen in the same year. Unfortunately, her material has disappeared from the University of London Library where it was deposited. Two points of detail are pertinent. First, the minutes of the Board of Studies in Geography for 29 November 1928 noted: 'Application for exemption from MSc examination, made by Miss Taylor was approved'. They also record the amended Regulations for Higher Doctorates (DLitt, DSc) as follows: 'The degree will be conferred in respect of printed and published work; in special cases, however, unpublished work may be taken into consideration'. Second, Eva wrote to Arthur Robert Hinks (1873–1945), then secretary of the Royal Geographical Society (RGS) and editor of its *Geographical Journal*, to request that her paper on Jean Retz might be accepted and published without delay: 'I am proposing to present my studies of Tudor Geography to the University as a thesis for a higher degree – and as it is a condition that work must be published, I hope you will find space for Jean Retz; he is only 3,000 words' (RGS Archives, E. G. R. Taylor Correspondence Block 1921–30, Letter from Taylor to Hinks, 6 February 1929). Three days later, she replied to Hinks:

> The position as regards the date of publication of portions of my thesis makes it simply 'as soon as possible' – I must get three or four papers published, and then group them together as 'Studies in Tudor Geography' – It is obvious that I cannot hope to get the complete thesis published as a whole, except as a book, and I believe the University Regulations will be satisfied by the fact that 'substantial portions' have been accepted for publication. (RGS Archives, E. G. R. Taylor Correspondence Block 1921–30, Letter from Taylor to Hinks, 9 February 1929)

Four days later, Hinks accepted the article subject to further discussion of details.

Eva's doctoral examiners were Edward Heawood (1864–1949), who had joined the staff of the RGS in 1884 and was its long-serving librarian (from 1901 to 1934) as well as being an expert on the watermarks of early modern printed papers, and Alexander Stevens (1896–1965), lecturer in geography at the University of Glasgow, whose early career had involved exploration in both Antarctica and in the Arctic, and work for the Ordnance Survey (Skelton 1951; Fairhurst 1966, 58; Robertson 1973, 15). Heawood and Stevens were well equipped to determine the quality of the work and it was not necessary to call upon the nominated reserve examiner, Dr Marion Newbigin (1869–1934) (*Geographers* Vol. 28). When Unstead retired in 1930, at the early age of 55 – in his own words, 'to read and think, to travel and write', Dr Taylor was appointed to succeed him in the chair of geography, in the face of several male contenders (cited by Campbell 1987, 45) (annual calendars for Birkbeck list Unstead as Director of Research in Regional Geography from 1930 to 1936). Taylor's fellow candidates for the Chair included Sidney William Wooldridge (1900–63), who had trained as a geologist at Kings College London, where he subsequently taught geography for almost all his career (on Wooldridge, see *Geographers* Vol. 8). Upon Eva Taylor's retirement in 1944, he would succeed her in the chair of geography at Birkbeck, only to return to Kings College in 1947 when a chair of geography was created in his home institution.

Eva Taylor was the first woman professor of geography in the British Isles. A further three decades elapsed before Alice Garnett (1903–89) at the University of Sheffield became the second female holder of a chair of geography in a British university in 1962 (Maddrell 2009, 192). As professor and head of department, Eva Taylor exercised considerable influence in the powerful Board of Studies in Geography, which administered and regulated every aspect of teaching and examining of the subject in the federal University. She was also a member of its Higher Degrees Sub-Committee in Geography, which approved, amended or disapproved areas of study and titles of research submitted for higher degrees, as well as appointing internal and external examiners for doctoral and Masters dissertations. The Minutes of the Board and Sub-Committee reveal that she attended assiduously but, despite unquestioned efficiency, she was never elected to chair such meetings. That function was undertaken between 1903 and 1929 by Edmund Johnston Garwood (1864–1949), who was professor of geology and physical geography at UCL from 1901 to 1931, and with Unstead as secretary during the 1920s, followed by Charles Bungay Fawcett (1883–1952) in the chair until 1946 (Clout 2003, 4) (on Fawcett, see *Geographers* Vol. 6).

The succinct nature of the handwritten minutes does not allow an appreciation of Eva Taylor's participation at these meetings. We do know, however, that she was outraged by decisions taken at a meeting she could not attend, and which affected two of her graduate students. She demanded that Fawcett convene a special meeting of the Sub-Committee at which she expressed her grievances at length. Fawcett declared that he and other members of the Sub-Committee had acted in the best interests of the students, advising one to slightly narrow the field of study and informing the other of overlap with research completed elsewhere. Professor Taylor insisted that each head of geography should be alerted in advance when issues relating to specific students were scheduled to arise and that a deputy be invited in the rare instances when a head of department was unable to be present to defend his or her corner. Two lined sheets of handwritten minutes, inserted into the normal minute book, reveal that members of the Sub-Committee acceded to Professor Taylor's request. Nonetheless, Fawcett proclaimed: 'The wording adopted by the University [clerks] in neither case embodied the Committee's recommendations exactly. He pointed out further that although the Committee normally endorsed the judgement of a "recognised teacher", it could not regard itself under any obligations to do so' (Archives of the University of London, AC 8/24/1/2. Minutes of the Board of Studies in Geography, 14 December 1938: the appellation 'recognised teacher' was specific to the University of London and related to a scholar who had passed probation and had satisfied the Board of Studies in terms of quality of teaching and productivity of research).

Following her elevation to the chair at Birkbeck College, Eva Taylor continued to publish prolifically during the 1930s, producing a second major book, *Late Tudor and Early Stuart Geography: 1583–1650*, editing volumes for the Hakluyt Society, and submitting an array of scholarly articles and book reviews (Taylor 1934a). Founded in 1846, the Hakluyt Society continues to publish works of travel and exploration that reveal the state of geographical knowledge in the past. Eva Taylor joined its Council in the early 1930s, and delivered a lecture on Richard Hakluyt at the RGS to commemorate the centenary of the Hakluyt Society (Taylor 1947a). During this period, she was one of the 39 founder members of the Institute of British Geographers, which came into being in 1933 as the professional body for university geographers, and as a mechanism for publishing geographical research. She belonged to its first editorial committee (1933–4) and she assisted in securing what was deemed to be 'a large grant' [£60] from the University of London to assist in the publication of Henri Baulig's monograph on *The Changing Sea Level* (on Baulig, see *Geographers* Vol. 4). She did not serve in any other capacity or publish in its *Transactions* (Baulig 1935; Steel 1984a, 15). She did attend its annual conferences, with several colleagues remembering that she enjoyed the company of younger members of the discipline (Maddrell 2009, 179). With its long-established concern for exploration, discovery and the history of map making, the RGS was the appropriate forum for her particular brand of research. She had been elected into its fellowship as early as February 1922, being proposed by Unstead and seconded by Edwin James Orford, a teacher and writer of textbooks. She was the second woman to be appointed to the Society's Council, serving from 1931 to 1935 and again from 1937 to 1941, and, as befitted her expertise, was a long-serving member of the maps and library committee. It was this commitment to the value of maps that caused her to make representations to the Director of the British Museum about the poor housing of the map collections of the Museum (Steel 1984a, 105). Throughout her career, she published many scholarly articles and book reviews in the *Geographical Journal*. Following an early initiative by the British Association for the Advancement of Science (BAAS), the RGS invited her in 1938 to assemble evidence on the distribution of industrial activity in England and Wales, which would be used in national planning during and after World War II. In 1947, the RGS would formally acknowledge her sustained commitment to the Society and to the discipline of geography by awarding her its Victoria Medal; she was further recognized when made an Honorary Fellow in 1965.

During the interwar years, Eva Taylor attended the meetings of the International Geographical Congress at Cambridge (1928), Warsaw (1934) and Amsterdam (1938) and presented research papers at each of them (Taylor 1934b). Like many geographers of the time, she was a strong supporter of the BAAS, being elected president of Section E (Geography) in 1939 and in 1947, when she delivered her address on 'Geography in War and Peace' (Taylor 1947b, 1948). The press turned to her for a pithy quote at these gatherings and one paper named her as 'Queen' of the BAAS (Maddrell 2009). Despite problems of ill-health, associated with increasing age and decades spent poring over books and manuscripts, she would in retirement write three research monographs, edit other works and deliver further scholarly articles, some in new journals. These included *Imago Mundi*, founded in 1935 to cover the history of cartography, which published five of her essays, and the *Journal of the Institute of Navigation*, established in 1946, in which she published over thirty articles.

Professor Taylor assembled a small team of geographers at Birkbeck College to teach the college's evening and weekend students. In 1930, she successfully recommended the appointment of Harry Cyril Knapp Henderson (1903–83), whose teaching ability in the field had impressed her during a fieldclass organized for the BAAS (Steel 1984b, 255). She subsequently supervised his doctoral research on 'The agricultural geography of the Adur Basin, in its regional setting', which was completed in 1935. The climatologist, Dr Hugh A. Matthews, who had trained at UCL, taught at Birkbeck during the 1930s as well as at Bedford College, the LSE and UCL. In 1931, Andrew C. O'Dell (1906–66), a graduate of Kings College London and of the LSE, joined Birkbeck College as assistant director and part-time lecturer in geography (Anon. 1967; East 1966). In 1933, Eva Taylor praised his research on the Shetland Isles, delivered to the RGS and subsequently published in the *Geographical Journal*; he was not made a full-time lecturer, however, until 1937 (O'Dell 1933, 514). With the exception of service in the Royal Air Force (Inter-Service Topographical Department) during the war, Henderson would spend the whole of his career at Birkbeck, becoming professor in

1965 and retiring in 1970. Matthews played an important role in the early years of the IBG (Steel 1984a, 12). Later, serving in the RAF, he died in 1943 at the age of 43 (Miller 1943, 63). During the war, O'Dell was drafted into naval intelligence work and then into planning research in Scotland, prior to leaving Birkbeck in 1945 for the University of Aberdeen where he became head of geography, and subsequently professor (on O'Dell, see *Geographers* Vol. 11). It was doubtless on his recommendation that Eva Taylor was made an honorary LLD of that university in 1948. Arthur Smailes, a graduate of UCL, also taught at Birkbeck College during 1938–9 (Anon. 1985, 149; Rawstron 1985, 120).

Although war deprived the college of Henderson, Matthews and O'Dell, Taylor was joined by Dr Edward Christie Willatts (1908–2000), a graduate of the LSE, who taught at Birkbeck in a part-time capacity from 1939 to 1942 (Worth 2000). During the 1930s, Willatts had been organizing secretary of the Land Utilisation Survey (LUS), orchestrated by Dr Laurence Dudley Stamp (1898–1966) at the London School of Economics (on Stamp, see *Geographers* Vol. 12). In 1941, Willatts was appointed senior research officer in the planning department of the Ministry of Works; seven years later he would become Principal Planner in the Department of the Environment, a post he would hold until retirement in 1973. Eva Taylor was not involved with the LUS but her colleagues Henderson (with doctoral student Eric Briault), O'Dell and, of course, Willatts were important contributors. Several other geographers at Birkbeck College produced county reports for the LUS: H. C. K. Henderson and E. W. H. Briault on Sussex (1942); H. C. K. Henderson (with A. N. Harris), Derbyshire (1941); A. C. O'Dell, Orkney (1939), Zetland (1940), Roxburghshire (1946), Fife (1946); E. C. Willatts (with L. D. Stamp), Middlesex and the London Region (1937), Surrey (1941).

With the assistance of Willatts and of her former student Alfred Edward Frank Moodie (1901–70), who was employed at Birkbeck from 1942 to 1944, as well as a current Masters student, Eila Campbell, appointed as part-time demonstrator and lecturer in 1941, Taylor continued to teach students during the war, concentrating her classes over weekends rather than during the week (Campbell 1973). Like several other geography academics, she also trained military officers in the arts of reading and interpreting European topographic maps, only leaving London during the worst phases of bombing when she retreated to her cottage at Bismore, near Stroud, in Gloucestershire (Maddrell 2008, 132). Indeed, hers was the only Geography department in the federal university not to be evacuated from central London, despite Breams Buildings suffering a direct hit during a bombing raid (Geographers from Kings College were evacuated to Bristol; those from UCL to Aberystwyth; and those from Bedford College, the LSE and Queen Mary were accommodated at Cambridge during the war).

In the autumn of 1944, Eva Taylor retired and became professor emerita. Despite health problems, including severe arthritis, sciatica and lumbago, she continued to lecture at Birkbeck College for some years following the appointment of Wooldridge as professor and head of department. She also attended professional meetings in central London, and pursued research with unflagging energy. This was facilitated by her acquisition of a *pied-à-terre* in Courtauld House at Byng Place, Bloomsbury, just a short distance north of the Reading Room of the British Museum. Her flat was notable for the paintings of nude women that adorned its walls, including 'Nature Unveiling Herself' by William Etty: her main home remained at Ralphs Ride, near Bracknell in Berkshire (De Clercq 2007, 5).

Several colleagues at UCL recall an occasion in the early 1950s, when Professor Henry Clifford Darby (1909–92) invited Eva Taylor to meet academic members and research students (on Darby, see *Geographers* Vol. 26). Hugh Prince remembers that she arrived from Byng Place by wheelchair, and W. R. Mead notes: 'It was a curious experience. She never said anything about her life, work or interests. She pointed her stick at one after another of the circle of seated geographers and said: "What do you do? What do you mean by that? Where will you publish?" My feeling was that it was a lost opportunity for her to say something about the origins of her own interests. That *would* have provided something for posterity' (W. R. Mead, pers. comm. 27 May 2011). However, the occasion enabled her to be the undivided centre of attraction for a group of young geographers (R. A. French, pers. comm. 2002).

In 1960, Eva Taylor was made one of the first Fellows of Birkbeck College for her distinguished contributions to knowledge and to the life of the institution. By this time, she was very frail and had come to depend on others in many ways. Eila Campbell was her unofficial assistant, acting as amanuensis, secretary, research assistant and errand runner, and making practical arrangements for her care and finances in the absence of her sons then working abroad (Wallis 1994; Prince 1995). In 1964, Taylor suffered a stroke that damaged an optic nerve and further hampered her already restricted mobility. Nonetheless, her intellect remained sharp and she expressed the view that the then controversial 'Vinland Map' was, in fact, a fake (Taylor 1974; Richey 2000). From her bed, she dictated her final articles to Eila Campbell, who became her academic successor both at Birkbeck College, when she became Professor in Geography in 1969, and in the world of the history of cartography. It was Campbell who saw Eva Taylor's final monograph, *The Mathematical Practitioners of Hanoverian England, 1714–1840,* through to publication (Taylor 1966). Professor Taylor spent the final years of her life at Saint Anne's Nursing Home in Wokingham, Berkshire, where she died of heart failure on 5 July 1966, aged 87.

Scientific Ideas and Publications

Between 1905 and 1930, Eva Taylor published a wide array of books and articles for schoolteachers and their pupils, starting with statistical appendixes for textbooks written by A. J. Herbertson for the Oxford University Press (a full bibliography of her works is to be found in Anon. 1968). In 1909, she collaborated with Herbertson on the Oxford Wall Map of North America, but by this time she was already working with Unstead on several titles that would be revised and kept continuously in print over the coming half century. The first of these was their *General and Regional Geography for Students,* which aimed 'to survey broadly the whole field of geography, and includes . . . physiography as well as political and economic geography, for only by the combination of these aspects of the subject can the study be either scientific or educational' (Unstead and Taylor 1910, v). This wide-ranging textbook was produced by George Philip and began a long and profitable association for the authors with this publishing house. Running to over 500 pages and comprising more than 100 maps and drawings (but with no photographs), it devoted 15 chapters to themes in world geography (including map reading, physical geography and human geography) before presenting a further 15 chapters on regions of the world. The authors explained: 'Britain, as the home region, has been dealt with in greatest detail; the space allotted to Europe and the other continents has been determined by their relative political and economic importance, but the portions of the British Empire situated in each region have received special attention' (Unstead and Taylor 1910, vi). The volume was adopted widely and reached its fifteenth, revised, edition in 1952, which took into account 'the great political and economic changes which occurred during and after the Second World War' (Unstead and Taylor 1952, vi).

Encouraged by this success, George Philip repackaged sections of their textbook as a *Geography of the British Isles* and as *Commercial Geography, General and Regional* (Unstead and Taylor 1911a, b). The two authors insisted:

> Unless the study of commercial geography is to be limited to a mere memorizing of place names and associated products, it must be based upon a sound knowledge of the outlines of physical geography. Once the student realizes that particular facts as regards relief, climate and vegetation are associated with particular consequences as regards industry and commerce, the work becomes not only more interesting but far more intelligent as the memory work is supplemented by the exercise of reason. (Unstead and Taylor 1911a, vii)

One reviewer of *Commercial Geography* declared: 'The chief merit of the book is that it certainly attempts to base commercial phenomena on the physical geography; but its claim to cater for examination candidates and to supply them with explanations and mental pictures is far from justified' [and] 'some of the statements are unquestionably misleading, if not absolutely incorrect' (Anon. 1912, 152). This allegation produced an angry two-page response from Unstead and Taylor to 'what must be regarded as ill-founded criticism' (Unstead and Taylor 1912a, 293). They quoted the sources for the 'unquestionably misleading' statements, and concluded: 'It is not . . . on questions of opinion that we mainly object to the reviewer's comments: the serious matter lies in his mis-statements and denial of the facts for which we have above given our authorities' (Unstead and Taylor 1912a, 294). Despite this early skirmish, the book sold well and reached its seventh edition in 1933, when a special version was also produced for South Africa that highlighted the geography of that territory.

In 1912, George Philip published *The Essentials of World Geography*, comprising 17 thematic chapters illustrated with 80 maps and line drawings (Unstead and Taylor 1912b). The authors explained: 'The chief physical and human factors are first considered separately, and then taken together so as to bring out the combined characteristics of each region of the world. In this way the natural regions of the globe are obtained' (Unstead and Taylor 1912b, v). This title proved yet another success and would reach its eighteenth edition in 1954. Between 1911 and 1916, Unstead and Taylor drafted *Philip's Comparative Series of Wall Atlases of the Continents*, with an explanatory handbook for each set of maps. In 1915, a further series, comprising foundation maps and transparent overlays, appeared as *Philip's Synthetic Maps*.

During 1915, Eva Taylor wrote three 30-page booklets for G. Philip, as *Questions and Exercises on Mackinder's Elementary Studies in Geography* (which had appeared as *Our Own Islands*, *Lands Beyond the Channel* and *Distant Lands*) (Taylor 1915a, b, c). Mackinder was well aware of the demands being made by his young colleague. In the preface to *Questions and Exercises on Lands Beyond the Channel* he explained:

> This series . . . is rather more advanced in character than the corresponding series upon *Our Own Islands*. The pupils must seek answers not only within the covers of the book itself; they must be prepared to turn to an Atlas of orographic maps, and to the books of the series they have previously read . . . They are required, too, to cast their answers occasionally in essay form, and here they will depend on the teacher for information supplementing the text. (Mackinder in Taylor 1915b, 3)

In the preface to the third booklet, he remarked: 'These *Questions and Exercises* . . . are of a more difficult type than the previous sets, and demand a greater independence of thought and judgment' (Mackinder in Taylor 1915c, 3). Among scores of questions, pupils were required to respond to 'How was geographical discovery influenced by the

Turkish invasion of south-eastern Europe?', and to write an essay on 'Man's debt to the winds'. Drawing on her experience in training colleges, Miss Taylor wrote six pamphlets that dealt with *The Teaching of Geography to Young Children*, and she published several short didactic articles in *The Geographical Teacher*, the then journal of the Geographical Association (Taylor 1916, 1920).

At this time, she produced several other titles, including *Long Ago and Far Away*, and *From Mountain to Sea* (Taylor 1919, 1922). Further collaboration with Unstead yielded *The British Isles and the Neighbouring Lands*, the 22 chapters of which drilled children in 'Learning from the map' and then discussed north-western Europe with special attention to Britain and Ireland (Unstead and Taylor 1919). The authors admitted: 'children of 10 to 11 years old, reading this book for the first time, may find parts of it rather hard. At this stage, however, the teacher expects to give a great deal of help'. They remarked sagely: 'It is a curious thing that we find it harder to describe our own country, of which we know and have seen so much, than to describe a foreign country of which we know little'. 'Why is Canterbury the seat of the Primate of the English Church?' figured among suggested exercises, alongside the instruction to 'Make a sketch-map to illustrate the coming of Saint Augustine from Rome' (Unstead and Taylor 1919, 78–9).

Much more innovative was Eva Taylor's *A Sketch-Map Geography: A Textbook of World and Regional Geography for the Upper and Middle School*, published by Methuen (Taylor 1921). As she explained:

> This book exactly reverses the procedure of the ordinary geography textbook. Instead of text illustrated by a few maps, it consists of a very large number of maps, supplemented by a terse series of explanatory notes. The maps are to be studied first. . . . Not one of the maps is a mere transcription from an atlas. In an atlas, facts as to topography, climate, vegetation, crops, and so forth are mapped separately. These sketch-maps show different categories of facts in combination, where they are seen to be, not merely facts, but factors, determining some particular geographical result . . . The educational value of the book is mainly in the opportunities it gives for scholars to work out for themselves causal relations and geographical principles . . . It also encourages the sketch-map habit. (Taylor 1921, v–vi)

Dealing first with regional geography, then world themes, and, finally, offering memory and revision aids in the form of 'shorthand sketch maps', the book went through 13 reprints before being completely revised in 1950 with the help of Eila Campbell, when all the maps were redrawn in a bolder, simpler style; a further revision was made a decade later (Taylor and Campbell 1962).

In 1923, Eva Taylor brought out *The Business Man's Geography* for George Philip, a book that went through three editions before being re-titled *Production and Trade. A Geographical Survey of all the Countries of the World* (Taylor 1930a). In his review, L. Dudley Stamp declared that this was not 'an entirely new compendium' as the publishers claimed: 'The plan of the book remains unchanged, the format, maps, type, paper, and other details show little alteration – all of which is acknowledged by the author in her preface. It seems a little curious that the publishers should thus tend to rob the author of the credit which naturally accrues from a public demand for a new edition and at the same time rather mislead purchasers' (Stamp 1931, 486). Treating the countries of the world alphabetically, the maximum of information was included in the minimum of space but, to Stamp's annoyance, dates for the statistics were rarely provided and there were no references to assist the business man or commercial student seeking further information. The reviewer for the Geographical Association complained that there were no tables or graphs of statistics, and noted that 'the maps are not the strongest feature of the book', while the lack of references was 'a serious omission' (S. E. J. B. 1931, 322). Similarly, Eva Taylor's *Oceans and Rivers* had received a chastening review from Marion Newbigin who concluded: 'It is not easy reading, and the attempt made in a few cases to put into words facts and deductions which can only be made clear by diagrams and figures illustrates needless complexities' (Taylor 1928; Newbigin 1929, 180). The fact that 'the bibliography does not seem to have had so much care given to it as some other parts of the text' was a matter of further regret (Newbigin 1929, 181).

By contrast, Miss Taylor's *Foundation Exercises in Geography: Simple Sketch Maps for Individual Work*, comprising a series of 32-page booklets for Philip met with great success (Taylor 1927). Their author declared severely: 'There is no avoiding it; every science must have a foundation, and that foundation is *fact*. But the acquisition of *facts* need not be burdensome. Children may be as eager in the pursuit of *facts* as to the pursuit of ideas. The fundamental *facts* of geography, upon which all geographical ideas, all geographical reasoning must be based are most economically expressed in sketch maps' (Taylor 1927, cover). A reviewer in the *Geographical Teacher* stated that this was 'an interesting set of booklets for small folk. If the exercises are used as indicated, the pupil cannot fail to obtain and retain definite, accurate information in relation to essential geographical text. More than this, he (*sic*) will probably face the matter of illustrative sketch maps not only with better grace, but also with considerably improved technique' (W. E. W. 1929–30, 63). From six initial titles, the series grew to nine as booklets on Wales, Scotland and West Africa were added. By the mid-1960s, some would reach their thirteenth edition.

In addition to these pamphlets, Eva Taylor was working on a much more ambitious project, nothing less than a two-volume *Complete School Geography* for Methuen, running to almost 800 pages, illustrated by 170 maps and diagrams, and claiming to be 'a complete preparation for all examinations of matriculation standard, and for commencing first-year university work' (Taylor 1926, xi). Emulating the basic structure of earlier work with Unstead, Miss Taylor discussed 'The World: its physical geography, peoples and products' in the first volume, and proceeded to explore 'The World: its political geography, industries and commerce' in the second. Unlike many of her other books, this title did not suggest exercises at the end of each chapter, but did include an appendix on 'Suggestions for study' in each volume. Distinctive features included: 'How to study the home district', and 'How to study a special region in detail'; a second edition appeared in 1936. The copies examined at the Institute of Education of the University of London have these two chapter headings underlined in pencil. By rich coincidence, the books belonged to Richard Lawton, who would go on to be professor of geography at the University of Liverpool. Lawton had been presented with the work in 1938 as a prize in the 'First Class "A" Division' at Houghton-le-Spring Secondary School in County Durham when he was in Form IIA.

Alongside this prodigious output of educational textbooks, Eva Taylor had by the mid-1930s published a suite of scholarly articles, achieved her doctorate, produced two research monographs and acquired her chair in the University of London. A turning point had not only been reached but passed, both in her career and in the establishment of geography as a university discipline. Writing textbooks for schoolchildren probably held little further attraction, but the chore of revision was doubtless sweetened by the regular arrival of royalties.

In August 1930, Eva Taylor put the finishing touches to her *Tudor Geography, 1485–1583*, which was published by Methuen, a company with which she had a good working relationship. She explained there how, following the European (re-)discovery of America:

> The world location of England had been completely altered: from her slumbers on a remote margin of the Old World, she had awakened to find herself on the very threshold of the New. Elizabeth's day saw the map and the globe as the necessary furniture of the closet of the scholar, merchant, noble and adventurer alike, and dreams of Empire were formulated which found expression in [Francis] Drake's achievement. [The end] of this study . . . was marked by the withdrawal of John Dee, the man behind the scenes of overseas enterprise; it saw, too, the firm establishment of the younger Richard Hakluyt as the foremost propagandist for expansion. His work as a geographer, and that of his emulator, Samuel Purchas, will form the theme of a later volume. (Taylor 1930b, v)

Drawing on her previous articles, Eva Taylor first offered a survey of English geographical literature prior to 1583, and then charted the role of Elizabethan entrepreneurs, voyagers, politicians and philosophers, with three of her eight chapters focusing on the activities of John Dee and the search for a route to Cathay. This narrative section of 160 pages concluded with a review of practical surveying and navigation, a theme that would assume greater significance in later work. The second section of the book presented the source materials used to construct the narrative: a catalogue of English geographical works, printed and manuscript to 1583, with notes; a list of the works of John Dee; and a list of illustrative and evidential documents. This division between message and supporting information provided a device that would structure much of Eva Taylor's later work.

Reviewers were generally positive in their appraisal of *Tudor Geography*. Gerald Roe Crone (1899–1982), librarian and map curator at the RGS, emphasized how Eva Taylor demonstrated that the Elizabethan voyages of discovery 'were carefully planned in accordance with current geographical conceptions, however erroneous these were subsequently proved to be. The moving spirit was Dr. John Dee, whose character as a serious geographer is here vindicated, and who is shown to have been the link between the "theorists" [he counted Mercator and Ortelius among his friends] and the explorers' (Crone 1931, 289; Anon. 1983). What Crone believed to be lacking from the book was a discussion of the processes of physical geography, and a clear statement of how the ideas of medieval scholars were being replaced by new conceptions based on the accumulation of observations. G. B. Parks, a historian at the Queens College of the City University of New York, declared the book to be 'good history as well as geography', and particularly appreciated the discussion of contemporary mathematicians and of John Dee (Parks 1931, 686). Fellow countryman, Francis Johnson, writing for *Isis*, the journal of the History of Science Society, concluded that the book was scholarly and 'by far the most important book yet published dealing with Elizabethan science' (Johnson 1935, 290). Following rapidly on this major work, Eva Taylor edited and introduced *A Brief Summe of Geography, by Roger Barlow* for the Hakluyt Society, beginning an association with that learned society that would extend across the rest of her life (Taylor 1932a).

In 1934, Professor Taylor delivered her second monograph to Methuen, entitled *Late Tudor and Early Stuart Geography, 1583–1650*. She ensured that Crone's criticism was met right at the start, declaring that this was the period 'in which truth is sought by experiment and observation instead of, as formerly, by reference to authority and revelation . . . The first half of the seventeenth century was the twilight that heralded the dawn of modern science . . . three English parsons, Richard Hakluyt, Samuel Purchas, and Peter Heylyn, spokesmen for successive reigns and generations, epitomised in turn the geography of their day' (Taylor 1934a, v). She also raised an important methodological point, which shifted substantially from her previous advocacy of facts, and she also defined what she meant by 'geography':

> It will be objected, of course, that many of the writers quoted are inaccurate and ill-informed. That is true, but it is true also of our modern daily press: each mirrors current thought and belief rather than current fact. And it is upon thoughts and beliefs, however faulty, that men act. The present writer's purpose is not to reconstruct the geography of the seventeenth century as it was, but as men believed it to be . . . The term 'geography' has designedly been interpreted in its widest sense, to include voyages and travels; maps and survey; mathematical geography and navigation; general, regional and descriptive geography; human, political and historical geography; physiography and economic geography. Certain works not in themselves geographical [such as sermons and plays] have also been included where they throw light on geographical conditions. (Taylor 1934a, vi–vii)

Twelve narrative chapters covering the first 175 pages discussed Richard Hakluyt, Samuel Purchas, voyages, chorography, mathematical geography, navigation and surveying, before tracing selected themes such as nature, human geography, urban travellers and colonial geography. The second part of the book, running to 120 pages, provided a bibliography of and index to nearly 2,000 manuscript and printed works relating to English contributions to geography prior to 1650.

Once again, Parks was largely enthusiastic, praising the wide scope of sources consulted, which ranged 'from drainage plans to reports of earthquakes, from atlases and travel collections to pamphlets of trade. The compiler's industry was prodigious, and the result admirable, [however] one must regret that the manuscript items are . . . left unlocated and thus in a manner re-interred' (Parks 1934, 693). The reviewer recommended further referencing to contemporary developments on the mainland of Europe, as 'it appears now that the history of geography in England can be understood only in relation to its history in Europe as a whole' (Parks 1934, 694). The Oxford geographer and historian of geography, John Norman Leonard Baker (1893–1971) congratulated Taylor on the book, which he thought better written than the preceding volume and 'deals with the subject in a more systematic and geographical manner', but bemoaned the lack of references, arguing that 'it would have been a small task to indicate, at least, in which library the manuscripts were to be found, and it would have helped considerably to be given more precise information about those which have been printed' (Baker 1934, 171–2) (on Baker, see *Geographers* Vol. 16). Nonetheless, he awaited a third volume from Professor Taylor on the history of English geography with eager anticipation. Francis Johnson found this book 'decidedly inferior to the earlier volume' and, after identifying various inaccuracies, was especially annoyed by the bibliographies, which he felt were 'practically useless to all other scholars' (Johnson 1935, 292–3). The journal of the Geographical Association was more generous, proclaiming: 'This is clearly pioneer work . . . Like its predecessor, the book is essential . . . as an aid to research and as a picture of the geography of the times' (A.D. 1933, 226). Soon after its appearance, Eva Taylor edited and introduced a two-volume work, *The Original Writings and Correspondence of the Two Richard Hakluyts* for the Hakluyt Society (Taylor 1935). Parks proclaimed the success of 'the historian of Tudor geography [who] has brought together for the first time the scattered treatises and letters (and other personalia) of two Elizabethan geographers' and has written 'a skilfully compact life of the Hakluyt cousins' (Parks 1936, 527).

The idea of 'a picture of the geography of the times' clearly guided Professor Taylor as she contributed two chapters to that edited volume, *An Historical Geography of England before AD 1800* (1936), which the young Cambridge geographer, H. C. Darby, was preparing for Cambridge University Press. Between the two world wars there was much debate about the relationship between the well-established discipline of history and the infant discipline of geography, with areas such as 'historical geography', the 'history of geography' and 'geographical history' giving rise to controversy (Darby 2002). On 6 January 1932, representatives of the Historical Association and the Geographical Association held a meeting at the London School of Economics chaired by Oxford historian John Linton Myres (1869–1954), an advocate of geography (*Geographers* Vol. 16). During the exchange of views, Taylor declared: 'The application of the adjective "historical" to the noun "geography" strictly speaking merely carries the geographer's studies into the past: his subject matter remains the same' (Taylor in Myres 1932, 42). Darby found this statement helpful for his research – while accepting that descriptive cross-sections for particular periods needed to be complemented by explanatory work on processes of change – and invited Eva Taylor to contribute to his edited volume (Darby 1987, 2002). Using descriptions by John Leland (*c.* 1503–52) and by William Camden (1551–1623) as her principal sources, Taylor wrote two 'pictures of the geography of England', one for the reign of Henry VIII, the other that of Elizabeth I.

Leland surveyed libraries and muniment rooms in the doomed monasteries and then, starting in 1539, embarked on five journeys to various parts of the kingdom, recording information from observation and discussion, as well as what he found in written sources. His voluminous notes were never converted into his proposed 'Description of the Realm of England', since he was afflicted with mental illness towards the end of his life (Taylor 1936a, 330). Being born about the time Leland died, Camden 'plundered his predecessor's manuscripts . . . for topographical detail', and recorded the contemporary scene 'somewhat perfunctorily' since his main interest was in Roman Britain. It was, nonetheless, 'possible to extract from his pages an outline geography of Elizabethan England, and to discern especially the elements of permanency in the rural scene' (Taylor 1936b, 354). Camden's great work, *Britannia,* was started in 1577 and published, in Latin, in 1586, with an English translation appearing later. Providing a county-by-county description, it was a work of chorography and explored landscape, geography, antiquarianism and history. It enabled Eva Taylor to write an account of the human geography of England during the reign of Elizabeth I, which was very different in content from the content of her monographs. In retirement, she would edit *An Atlas of Tudor England* for Penguin, which used plates from John Speed's pocket atlas (Taylor 1951).

Taylor's enquiries into geography in early Georgian England did not give rise to a research volume, as new demands were placed on her time during the late 1930s, and her interests shifted towards the history of science, mathematics and navigation (Crone 1966, 595). Her next major book, *The Mathematical Practitioners of Tudor and Stuart England,* did not appear until 1954, ten years into her retirement (Taylor 1954). In the preface, the Astronomer Royal H. Spencer Jones, explained that 'navigation up to and even beyond the sixteenth century was almost entirely a practical art, in which success depended upon experience, sound common sense and good seamanship' (Jones in Taylor 1954, ix). A complete separation existed between seamen, who were ignorant of mathematics, and mathematicians, who had no practical experience of sailing.

> It was only gradually that a demand for instruction on geometry and astronomy arose to meet the need for improved techniques in navigation, surveying, gunnery, and horology. This need was supplied in some measure, though inadequately, by

> the mathematical practitioners of London before it began to be provided by the universities. Many of these teachers were men with practical experience but no great education, who worked in close association with the instrument makers. (Jones in Taylor 1954, x)

Taylor advised general readers to focus on the nine chapters of narrative (160 pages) that were arranged chronologically, while 'the student of the history of science' was recommended to explore the subsequent 270 pages that presented biographies of the practitioners, and an annotated list of contemporary works on the mathematical arts (Taylor 1954, xi). Derek Price, of the Whipple Museum of the History of Science in Cambridge, stressed: 'For the first time, it can be seen how enormous is the aggregate of the efforts . . . of those minor characters on the stage of science . . . and how much has been lacking from previous histories' (Price 1955, 16). Yet, as with her previous books, Francis Johnson complained: 'By failing to cite the sources of her information, the author greatly detracts from the value to other scholars of her compilations . . . [thereby] making her book distressingly unsatisfactory for later students who rightly desire to make it the starting point for their own researches' (Johnson 1957, 377–8). Crone explained that this monograph centred on 'the use of geometrical instruments for precise measurements in practical affairs' and was really 'a work for the specialist' (Crone 1955, 104). Although welcoming Taylor's map that showed the workplaces of mathematical practitioners in London as 'a thoughtful addition', he maintained 'complete reserve as to its origin', and complained too that it lacked a scale (Crone 1955, 105).

In fact, Eva Taylor was preparing a much more accessible book, entitled *The Haven-Finding Art: A History of Navigation from Odysseus to Captain Cook* (Taylor 1956). Michael Richey, executive secretary of the Institute of Navigation, was thanked for his inspiration and editorial skills, with this acknowledgement confirming that her intellectual voyage had moved beyond conventional geography. Illustrated with 28 figures and 25 plates, the book dispelled the persistent myth that early sailors navigated by 'hugging the shore'. In the foreword, the Hydrographer of the Navy exclaimed: 'Nothing is more fraught with peril, and therefore the more assiduously avoided on a little-known coast, than hugging the shore. The myth is based on the assumption that the mariner had neither the means nor the ability to find his way out of sight of land. Now this assumption is shown to be groundless' (Collins in Taylor 1956, x). This clarification was achieved in five readable sections working from how early sailors read 'signs in the sky' and changes in the surface of the sea; through navigation without magnetic compass or chart; to sailing with compass and chart; the subsequent use of instruments and tables; and ultimately progress towards 'mathematical navigation' whereby the challenge of determining longitude at sea was achieved. G. P. B. Naish, of the National Maritime Museum, was full of praise for this 'model of packed information, narrated clearly and amusingly, with just the right amount of reference to original sources, [that] closes with the successful invention of the marine timekeeper [chronometer] and the brilliant navigations of Captain Cook' (Naish 1956, 85).

Despite advancing years and increasing infirmity, Eva Taylor continued to write about cartography, survey and navigation, and edited two further volumes for the Hakluyt Society, *The Troublesome Voyage of Captain Edward Fenton, 1582–1583*, and *A Regiment for the Sea and Other Writings on Navigation by William Bourne of Gravesend, a Gunner, ca. 1535–1582*, and worked on another research monograph (Taylor, Singer and Price 1957; Taylor 1959, 1963a; Williamson 1960). C. H. Cotter reported: 'It is a vivid memory of . . . seeing the elderly lady, regularly and frequently, studying acutely in the library of Birkbeck College during the evening hours when this famous institution was (and still is) a hive of scholarly activity' (Cotter 1967, 99). Two books were the result. The first was a slim, well-illustrated text entitled *The Geometrical Seaman* (written with Michael Richey), which presented early nautical instruments (Taylor and Richey 1962). The second was the substantial *Mathematical Practitioners of Hanoverian England, 1714–1840*, also assembled with the help of Richey, the staff of the Institute of Navigation, and academic colleagues, especially Eila Campbell (Taylor 1966). Adapting the conventional structure, Eva Taylor first related the story of navigation from the accession of George I to 1840, a time when 'colonial expansion and commercial enterprises had led to a vast increase in maritime adventures, but the problem of finding one's way across the high seas was a very pressing one' (Scott 1967, 300). In 1714, Parliament had 'offered a huge prize [£20,000] for the determination of longitude at sea, making the first part of the eighteenth century a veritable whirlpool of ideas. But ideas were not enough, [as there was] a vast difference between basic principles and their practical realization' (Sadler in Taylor 1966, v). Taylor then presented biographies of some 2,000 mathematical practitioners, greatly varied in 'rank, education, calling, methods of thought and expression; whether they were thinkers or doers, great or humble, they all had in common a practical attitude toward their problems, loosely associated with the mathematical approach' (Sadler in Taylor 1966, v).

In reviewing the book, H. R. Calvert, Keeper of Astronomy and Geophysics at the Science Museum in South Kensington, praised 'the graphic style of the narrative [that] gives the impression that Professor Taylor lived amongst the many people she writes about. The result can only have been achieved by thoroughly absorbing most of the relevant literature of the period' (Calvert 1966, 520). There were, however, shortcomings: Calvert found that the one biographical entry that he checked thoroughly contained five errors of fact, presumably because the author's ill-health precluded the identification of transcription errors from her hand-written notes. Calvert concluded:

> Professor Taylor tried to make her book appeal to both the growing body of students of the history of science, and to those who need a reference book on the subject. For the former, there are too many unimportant details, and for the latter no easy way of discovering the source of any particular statement. This would not matter so much if the statements were all reliable, but there are far too many errors and slips. (Calvert 1966, 521)

One specialist went so far as to declare that consulting the book should come with 'a government health warning' (Darius 1990, 31). Despite its deficiencies, the book was reproduced in 1989 by an American dealer of rare books, as was *The Mathematical Practitioners of Tudor and Stuart England*, which had been reprinted three times during the 1960s (De Clercq 2007, 10). A demand for these monographs continued, but more accurate and sophisticated works on early modern British mathematicians, mathematical practitioners, instrument makers, land surveyors and map makers have been completed since the death of Eva Taylor (De Clerq 2007, 11).

An important aspect of the latter years of Taylor's university career was her contribution to the war effort between 1939 and 1945, when economic and spatial planning became a matter of national necessity. As early as 1929, at the start of the great depression, Eva Taylor had chaired a committee appointed by the BAAS to prepare a scheme for a National Atlas, which she would advocate on many occasions in later years (Taylor 1939, 1940a, b). Public concern over regional disparities, evidenced by mass unemployment in the old, often coal-based industrial heartlands and economic buoyancy in and around London, resulted in the appointment in 1937 of the Royal Commission on the Distribution of Industrial Population under the chairmanship of Sir Anderson Montague-Barlow (popularly known as the 'Barlow Commission') (Willatts 1971, 312). The RGS, among other bodies, was requested to give evidence and its council agreed to prepare a memorandum. Despite approaching 60 years of age, and having a heavy burden of teaching and administration, Taylor agreed to convene a committee of Fellows and to prepare the required memorandum, despite having no cartographic or technical assistance at Birkbeck College. It had been anticipated that the task would be shared with Dr L. Dudley Stamp, but since he was already committed to work in India at that time the burden fell entirely on Professor Taylor: 'Within six months, with voluntary help from university departments [in various British universities], she had drafted a memorandum of evidence and a portfolio of forty-nine maps. The latter, she said, "are the essential part of the evidence, for they reveal relationships and suggest guiding principles which do not emerge from tables of statistics or verbal memoranda"' (Willatts 1971, 312; Willatts 1987, 104). The geographers who contributed information were S. H. Beaver, K. C. Edwards, C. F. W. R. Gullick, H. C. K. Henderson, G. Manley, A. C. O'Dell, W. Smith, E. C. Willatts and S. W. Wooldridge (Campbell 1987, 51).

Following negotiation over the group's leadership by Arthur Hinks when Stamp returned, she convened an open meeting at the RGS at which the evidence was subjected to critical scrutiny (RGS Archives, E. G. R. Taylor Correspondence Block 1931–40, Letter from Hinks to Taylor, 29 April 1938; Taylor 1938, 22–7). She and Dudley Stamp gave oral evidence in November 1938 to members of the Commission who were impressed by the cartographic demonstration of spatial relationships between regional variations in sunshine and rainfall on health, housing and amenity. Through use of a device of 'sieve maps' to mask out or to highlight phenomena, Taylor's graphics revealed an 'axial belt' running from Greater London through the Midlands to South Lancashire and the West Riding of Yorkshire, within which there was a high degree of accessibility by road and rail, and there were major pools of skilled and semi-skilled labour. Whether this axial belt was a 'doctrine', a 'theory' or simply a factual generalization provoked lively debate among geographers. Eva Taylor declared it was simply the latter and withdrew – presumably in exasperation – from one meeting. John N. L. Baker regretted her departure, commenting at her expense, 'I am only sorry that she has gone before I had the opportunity of pointing out that she makes the best of both worlds; for part of the time in the axial belt [in London], and for part of the time in the delightful country of rural England', doubtless a reference to her cottage in Gloucestershire (Baker and Gilbert 1944, 69). Popular reports described this axial belt as 'Professor Taylor's coffin' following Sir Montague's memorable declaration that it was coffin-shaped, and others likened its morphology to that of an hour-glass. The RGS memorandum concluded as follows: 'The fundamental question that has to be decided is whether industry is to be forced or cajoled back into the old distributional pattern, or whether the industrial population is to be assisted to adjust itself to a new. With that problem, however, a Geographical Society has nothing to do' (cited in Willatts 1971, 313). Eva Taylor believed that geographers should investigate matters of social relevance, but she also believed that matters of policy were the responsibility of Parliament and politicians, rather than of academics and learned societies.

Although completed in August 1939, the 'Barlow Report' was not presented before Parliament until early in the next year. The majority of the Commission recommended that a national authority should be set up to research, advise upon and regulate the location of industrial activity. Eva Taylor again argued the importance of creating a national atlas to assist this proposed national authority, delivering lectures and writing articles to argue her case. During 1940, she was appointed to the national Panel of Reconstruction, and in the following year to the Consultative Panel of the Ministry of Works and Buildings. Early in 1941, the Minister of Works was given responsibility for reconstruction of bombed towns and cities, and for spatial planning (Taylor 1941, 1942, 1943). An advisory Maps Committee, including Taylor and Stamp, made a detailed proposal for a loose-leaf atlas of maps to assist future planning activity. A special cartographic office was funded and the first ten sheets were printed by 1944.

During and immediately after the war, Taylor was actively interested in the work of the Association for Planning and Regional Reconstruction, and helped that body to assemble evidence to submit to the committee on the future utilization of land in rural areas. Between 1943 and 1947, successive legislation reinforced the notion of planning and set up machinery to put it into motion. The Town and Country Planning Act (1947) established a comprehensive and mandatory system of spatial planning, which required a new type of planner with wider training than the conventional background in architecture. Eva Taylor, together with Stamp and Wooldridge, submitted evidence to the Schuster Committee on the Qualifications of Planners, appointed in 1948, which quoted her view that planning had now become 'primarily a social and economic activity, limited but not determined by the possibilities of design' (Willatts 1971, 315; Freeman 1980b, 59). The Town Planning Institute changed its examination for those seeking professional qualification to include a paper covering economic geography and applied geology. Having already co-drafted *Ground Plan of Britain: Fourteen Maps for the Use of Planning Activities and Groups*, Professor Taylor delivered *Geography of an Air Age*, produced *Planning Prospect* as the first part of a 'Survey before Plan' series, and published an article on 'The geographical basis of a county plan' in the *Journal of the Town Planning Institute* (Taylor and Balfour 1942; Taylor 1945a, b, 1949a).

On 21 January 1946, the Research Committee of the RGS invited her, in her absence, to draft a Memorandum on Post-War Needs in Geography, arguing that its members knew 'of no geographer who is likely to produce a better one' (RGS Archives, E. G. R. Taylor Correspondence Block 1941–50, Letter from Secretary of the RGS to Taylor, 1 February 1946). With some reluctance, she agreed, noting that 'the task allotted to me is a punishment for absence' (RGS Archives, E. G. R. Taylor Correspondence Block 1941–50, Letter from Taylor to Kirwan, 5 February 1946). She prepared her report over the next two months and was furious when it was dismissed by the RGS. She duly informed the Director: 'After spending many hours of thought and work on the Memorandum . . . I do not receive even a conventional "thank you", but simply see the thing dropped into the W.P.B. [waste paper basket], it is not encouraging' (RGS Archives, E. G. R. Taylor Correspondence Block 1941–50, Letter from Taylor to Kirwan, 21 March 1946). By this time, however, her main interests had shifted to what Walter Freeman described as 'fascinating if somewhat esoteric studies' in the history of mathematical knowledge, navigation and science, rather than academic geography, the history of geography or the challenge of national planning (Freeman 1980a, 129).

Influence and Spread of Ideas

Now a neglected name among British geographers, Eva Taylor exercised important influence on the subject during her lifetime. Her textbooks conveyed a rigorous and holistic vision of geography to pupils and schoolteachers for over 50 years. The lectures she delivered at Birkbeck College for a quarter century shaped the careers of hundreds of undergraduates, many of whom entered or advanced in the teaching profession. For the most part scholarship within British historical geography shifted to Britain itself and Domesday, marginalizing Taylor's huge body of work and conceptual developments in the field. Her later monographs were more fully acknowledged in the domain of the history of science and of navigation, while her interests in the history of survey and cartography have been appropriated by historians rather than geographers. As a professor in the federal University of London, she held a seat on the influential Board of Studies in Geography where she expressed her views during – and after – her tenure, when she was invited to attend meetings until January 1945 (Archives of the University of London, AC 8/24/1/3. Minutes of the Board of Studies in Geography, 26 October 1944). For reasons that were not documented, she continued to attend regularly until 1958.

At Birkbeck College, she mentored scores of advanced students who earned Masters degrees by research as well as by examination, and she supervised PhD students on a surprisingly wide range of topics at a time when doctoral qualifications were rare. As well as H. C. K. Henderson, among her early doctoral students were E. H. Selwood (1932), who classified English communities using information from the 1921 census, and H. Thomas (1935), who wrote a historic-geographical survey of the alkali industry of Great Britain. In 1939, S. M. Ali analysed the geography of the Ghaggar plain in northern India, and E. W. H. Briault completed a study of the historical geography of part of east Sussex, before entering an extremely influential career as the chief Education Officer for the London County Council (subsequently of the Inner London Education Authority) (Bevan 1996). Professor Taylor's final two doctoral students both entered university careers, with A. E. F. Moodie's analysis of the political problems associated with the Italo-Jugoslav frontier (1942) being followed by a chair in the United States, and the discussion of the political geography of Malaya (1945) by E. H. G. Dobby (1906–81) reflecting his experience gained while teaching at Raffles College (later the University of Singapore), where he would become the first professor of geography (Campbell 1973). Taylor also exercised her influence on the profession through her book reviews, being a scrupulous and sometimes exacting critic, who condemned weak conceptualization and loose expression. Harsh words, spoken or written, undoubtedly did not endear her to some members of the small community of academic geographers, and could have a detrimental effect on careers. Editors sometimes managed to persuade her to moderate the tone of her judgement, but they did not always succeed. Among numerous examples, William Vogt's *Road to Survival* was criticized for 'its exaggeration and inconsistencies, its apparent disregard of human values', and Grenfell Price's *Western Invasion of the Pacific and its Continents* was condemned for 'having been rather casually put together from [his] commonplace book' (Taylor 1949b, 93; Taylor 1963b, 537). In review of Josué de Castro, of the Food and Agricultural Organization, Taylor confronted readers of his *Geography of Hunger* with 'exaggerations and mis-statements that border on the grotesque, [so that] the reader is inclined not only to dismiss the book as nonsense but to dismiss also from his mind the very real problems with which it is concerned' (Taylor 1952, 348).

There is a real irony in her criticism of R. E. Dickinson and O. J. R. Howarth, authors of *The Making of Geography*, for supplying 'bibliographic references . . . with a regrettable lack of scholarly care, and often in such a way as to be of little use to the student' (Taylor 1934c, 66). Reviewing his *Urban Geography*, Eva Taylor identified 'some fine confused reading in Professor [Griffith] Taylor's new book; the confusion to be explained, perhaps, by the fact that ideas chase one another so fast across his mind that he has no chance to pin them down and examine their validity' (Taylor 1949c, 100). A suggestion from the assistant editor of the *Geographical Journal* that she might moderate her review of *The Pastoral Industries of New Zealand* (1935), written by a colleague in the University of London, produced the following response: 'I am afraid that you must send Dr. [R.O.] Buchanan's book to someone else, as on learning that my review did not suit you I destroyed it. I'm afraid I don't hold your committee's view that serious blemishes are immaterial if a book is to be placed before an inexperienced reader! Many apologies for not writing before – but I might have said something really harsh!' (RGS Archives, E. G. R. Taylor Correspondence Block 1931–40,

Postcard from Taylor to Assistant Editor, 30 September 1935). Taylor's entirely positive review of Emmanuel de Martonne's two volumes on *L'Europe Centrale* in the *Géographie Universelle* series also contained a very emotive attack on her British colleagues (on de Martonne, see *Geographers* Vol. 12). On receiving the review, Arthur Hinks, editor of the *Geographical Journal*, noted:

> I have been looking with some doubt at a passage of your review of De Martonne, which I should like to suggest might be eliminated from the review. I fully agree that English professional geographers have not in the last ten years produced works of the high quality that might be expected of them; but is it because the professorships have been filled by explorers, surveyors, and geodetic surveyors? I cannot think so. Rudmose Brown and Debenham did during early years take some part in exploration, but that can hardly be a disadvantage to them in teaching geography, and I should hesitate to print anything in the Journal that suggests that the electors to university positions in the last ten years have been putting their money on the wrong horses. (RGS Archives, E. G. R. Taylor Correspondence Block 1921–30, Letter from Hinks to Taylor, 3 March 1932)

The reply to Hinks, posted two days later, declared:

> Of course as Editor you must cut out what you choose, but I think it a fundamental mistake to equate 'criticism' with 'reflection upon'; vigorous criticism is a most valuable stimulant. I should be the last to deny the importance of exploration, and the even greater importance of survey, but what I still insist is that . . . enormous masses of uncoordinated, and unsystematised material are brought home by men who are trained only as surveyors. Sustained, co-ordinated research into particular areas and problems and sustained effort to coordinate material are the paths of progress . . . not merely looking for something 'new'. This viewpoint needs vigorously pursuing in so far as geography is concerned, and I am doing my best in reviews and elsewhere (RGS Archives, E. G. R. Taylor Correspondence Block 1921–30, Letter from Taylor to Hinks, 5 March 1932)

The core of Eva Taylor's published review reads as follows:

> Professor de Martonne's masterly survey of the geography of Central Europe . . . indicates him as among the most able, the most lucid, the most convincing exponents of geography that the present century has produced. The reflections of an English reviewer when handling these splendid volumes cannot be happy ones. Here are flawlessly printed pages, enriched with double-paged maps in colour, really beautiful photogravure plates, and illuminating text-figures the execution of which leaves nothing to be desired, the whole at the low price of 230 francs. Where is the English publisher, the English printer, the English draughtsman and blockmaker who could produce such a work at such a price? And, more important still, where is the English geographer who could match M. de Martonne in range of knowledge combined with power of geographical synthesis? (Taylor 1932b, 60)

A more extreme example of 'vigorous criticism' appeared in Taylor's review of *Man's Adaptation of Nature: Studies of the Cultural Landscape* by Patrick William Bryan (1885–1968) (Bryan 1933). The author had gained a PhD from the University of London in 1924 for a thesis on the coal and iron industries of the United States. This was subsequently recast for inclusion in the textbook, *North America*, co-authored with Llewellyn Rodwell Jones (1881–1947) of the LSE (Jones and Bryan 1924) (on Rodwell Jones, see *Geographers*

Vol. 4). Since 1922, Bryan had taught geography at University College, Leicester, where he founded the departments of commerce and geography. In January 1933, he had submitted *Man's Adaptation of Nature* for examination for the award of a DSc, with Hugh Robert Mill (1861–1950) and Marion Newbigin being appointed as examiners (Archives of the University of London, AC 8/24/1/2. Minutes of the Board of Studies in Geography, 24 January 1933) (on Mill, see *Geographers* Vol. 1). Eva Taylor drafted a lengthy critique of *Man's Adaptation* for publication in the *Geographical Journal*. This appraisal worried Arthur Hinks, who asked her to explain her negative opinion, which she did in a subsequent letter (RGS Archives, E. G. R. Taylor Correspondence Block 1931–40, Letter from Taylor to Hinks, 3 March 1933). The review was duly published, in which Eva Taylor declared that Bryan had 'invented a jargon and proclaimed a philosophy', and that his reading and interpretation of the works of Humboldt, Ritter and Ratzel was partial and faulty (Taylor 1933, 352). There was, she argued, 'nothing new, apart from detail and nomenclature, in [his] use of the "cultural landscape" as the concrete mode of approach to human geography'. Precision of language as 'the essence of scholarly work' was absent, indeed, 'confusion of thought, and therefore of language, is apparent again and again' (Taylor 1933, 353). Wounded by this attack, Bryan sent a six-page, single-spaced reaction to Hinks, which provoked a vigorous defence from Taylor, and a further reaction from Bryan (RGS Archives, P. W. Bryan Correspondence Block 1931–40, Letter from Bryan to Hinks, 15 April 1933; Letter from Taylor to Hinks, 28 April 1933; Letters from Bryan to Hinks, 25 April 1933, and 12 June 1933). After many harsh words had been written, Hinks managed to calm the aggrieved author and the matter was dropped. Patrick Bryan was not awarded a DSc and remained at Leicester. Eva Taylor may well have been right in her appraisal of his work, but she too was sensitive to adverse criticism (Maddrell 2009, 177).

Taylor's work was of great interest to practical sailors as well as scholars of navigation and exploration. Novelist and sailor, Arthur Ransome, praised her *Haven Finding Art* in the *Times* and after reading her work on William Bourne, Francis Chichester, world-renowned sailor, was compelled to write to Taylor in appreciation of her knowledge and vibrant prose: '. . . Not only is all technical navigational description so lucid and interesting but you make the reader feel that he is right on the spot at the time, taking part in what is going on down the Thames' (Maddrell 2009, 185). Not long after her death, she was described as 'perhaps the greatest authority on the history of navigation', although she was not a sailor herself (Deacon and Deacon 1969, 33; De Clercq 2007, 8). To mark her eightieth birthday, an appeal was launched for funds to be raised for a lecture to be delivered every couple of years by a distinguished scholar in one of the branches of knowledge to which she contributed. Birkbeck College, the RGS, the Institute of British Geographers, the Hakluyt Society, the British Society for the History of Science and the Institute of Navigation collaborated in this venture. Over 200 colleagues subscribed to the fund, indicative of their respect for her. The first lecture in her honour was delivered in 1960, and the series continues to this day.

Taylor's research would perhaps have had more long-term influence on the discipline had she compelled others to follow her into historical research. Like others engaged in war work, full or part-time, personal research was put on hold between 1939 and 1945. While her research publications culminated during her almost 30 years of an extraordinarily productive retirement, this was not fed back into the discipline through teaching and postgraduate supervision. The focus of the post-war generation of historical geographers was centred on terrestrial Britain and the Domesday records, which made little or no reference to her research. Eila Cambell's work bridged both camps and Taylor would have been unable to continue to publish in the later stages of her life without her significant support. Campbell's failure to complete commissioned biographical pieces on Taylor only served to further obscure the significance of Taylor's ground-breaking work in the histories of geographical thought and practice, but numerous subsequent histories of the discipline also failed on that count (Maddrell 2009).

Conclusion

Eva Taylor was a formidably energetic scholar who was productive to the very last. As the first female professor of geography in a university in the British Isles, she succeeded not only in a man's world but also as an unmarried mother with two children. Indeed, she 'met the geographical establishment head on – and used it for her own purposes', justifying her choice to switch from chemistry, where she thought there were few opportunities for women, to geography (Maddrell 2009, 186). Early in her career, she wrote textbooks and pamphlets that influenced generations of pupils and schoolteachers, and applied herself to a gruelling routine of archival and library work that yielded major contributions to the history of geography in England. She adhered to very high academic standards of argument and logic that she expected to find in the work and writings of fellow scholars. Her incisive reviews show that she was often disappointed by what she read, but her own publications were not without their shortcomings, notably inadequate referencing. In the years surrounding World War II, she turned her no-nonsense, business-like approach to collating geographical information that would prove fundamental for the introduction of a national planning system. In retirement, she wrote further monographs, but the focus of her work had shifted to the history of science and navigation, and it is in those fields that her books are still remembered and consulted. Taylor's working library was incorporated into the general collection of the Library of Birkbeck College. Her personal collection of early printed maps is to be found in the Special Collections of the University of London Library (Senate House), and her research papers are housed in the British Library.

Eva Taylor was a very fluent writer who, in the opinion of Michael Richey of the Institute of Navigation, 'would never put pen to paper until she was absolutely certain of what she wanted to say and, with her clarity of mind and wonderful mastery of the language, this gave the writing enormous vigour . . . She combined a mastery of technical detail with a warm interest in the personalities involved, so that the man as well as his maps emerged from her researches' (Richey 2000, 196). She was a brilliant lecturer, with remarkable clarity of thought and expression, which delighted and convinced her audiences. She was blessed with a lively and stimulating mind and with many talents, of which tact was not one. As Gerald Roe Crone, with whom she had sparred, recalled, 'With her strongly held opinions, often provocatively expressed, and certain prejudices, she was sometimes a difficult conversationalist, but her evident determination to get to the root of the matter and her practical common sense usually carried the day . . . Her contempt for lazy thinking, her persistence in working for the accurate answer, and her wide knowledge are not likely to be forgotten' (Crone 1966, 596). Similarly, Raleigh Ashlin ('Peter') Skelton (1906–70), Superintendent of the Map Room of the British Museum, stressed that 'her intellectual honesty, alertness of mind and intolerance of humbug enabled her to illuminate practically every subject which she touched' (Skelton 1968, 115). Robert Steel found her 'a formidable person by any standards' (Steel 1984a, 30). Her pupil and friend, Eila Campbell, acknowledged: 'Eva Taylor was not universally popular among her contemporaries. She combined a sparkling wit with an uncanny gift of putting her finger on the weakness in anyone else's argument, and a tactlessness which endeared her to some and alienated others. Her reviews of the books of others were often vitriolic' (Campbell 1987, 55). In a more benevolent tone, T. W. Freeman remembered her as 'ever young in heart and mind', while W. R. Mead recalled her in gendered terms as 'a model Tudor chatelaine, perhaps better still, a stalwart Elizabethan matriarch. In the right costume, she would fit happily into the Tudor gallery at the National Portrait Gallery' (Freeman 1980b, 39; Mead cited by De Clercq 2007, 12).

Eva Taylor produced a prodigious body of research scholarship and school texts, leaving a huge geographical and interdisciplinary textual legacy. As a woman she crossed a number of previously gendered thresholds, notably in being the first woman to be appointed to a professorial chair in geography and to other learned bodies. The continuing biennial E. G. R. Taylor lecture is an echo of her impact on the discipline of geography and beyond.

Acknowledgements

We express our thanks to R. A. French, W. R. Mead and H. C. Prince for their recollections of Professor Taylor, and to archivists and librarians at the University of London Library (Senate House), the Institute of Education (University of London), Birkbeck College, the British Library and the RGS.

Bibliography and Sources

1. REFERENCES TO THE LIFE AND WORK OF EVA TAYLOR AND HER MILIEU

Anon. (1929), 'Obituary: William Henry Barker', *Bulletin of the John Rylands Library* 13, 225–6.

Anon. (1966), 'Obituary: John Frederick Unstead', *Transactions of the Institute of British Geographers* 38, 199–200.

Anon. (1967), 'Obituary: Andrew Charles O'Dell', *Transactions of the Institute of British Geographers* 42, 189–92.

Anon. (1968), 'Obituary: Eva Germaine Rimington Taylor', *Transactions of the Institute of British Geographers* 45, 181–6.

Anon. (1983), 'Obituary: Gerald Roe Crone', *Geographical Journal* 149, 270–3.

Anon. (1985), 'Arthur Eltringham Smailes, 1911–1984', *Geographical Journal* 151, 149–51.

Baker, J. N. L. and Gilbert, E. W. (1944), 'The doctrine of an axial belt', *Geographical Journal* 103, 49–72.

Baulig, H. (1935), *The Changing Sea Level*. London: Institute of British Geographers.

Bevan, J. (1996), 'Obituary: Eric Briault', *The Independent*, 9 March.

Bryan, P. W. (1933), *Man's Adaptation to Nature: Studies of the Cultural Landscape*. London: University of London Press.

Campbell, E. M. J. (1972) (revised by E. Baigent in 2004) 'Taylor, Eva Germaine Rivington, 1869–1966', *Oxford Dictionary of National Biography*.

Campbell, E. M. J. (1973), 'Alfred Edward Frank Moodie, 1901–1970', *Transactions of the Institute of British Geographers* 59, 159–61.

Campbell, E. M. J. (1987), 'Geography at Birkbeck College, University of London, with particular reference to J.F. Unstead and E.G.R. Taylor', in R. W. Steel (ed.), *British Geography 1918–1945*. Cambridge: Cambridge University Press, 45–57.

Clout, H. (2003), *Geography at University College London: A Brief History*. London: University College London.

Cotter, C. H. (1967), 'A historical geographer', *Journal of the Institute of Navigation* 20, 98–9.

Crone, G. R. (1966), 'Professor E.G.R. Taylor', *Geographical Journal* 132, 594–6.

Crowther, J. G. (1967), 'Eva G.R. Taylor: an appreciation', *Journal of the Institute of Navigation* 20, 94–101.

Darby, H. C. (ed.) (1973), *A New Historical Geography of England*. Cambridge: Cambridge University Press.

Darby, H. C. (1987), 'On the writing of historical geography, 1918–1945', in R. W. Steel (ed.), *British Geography 1918–1945*. Cambridge: Cambridge University Press, 117–37.

Darby, H. C. (2002), *The Relations of History and Geography*. Exeter: University of Exeter Press.

Deacon, G. E. R. and Deacon, M. (1969), 'Captain Cook as a navigator', *Notes and Records of the Royal Society of London* 24, 33–42.

De Clercq, P. (2007), 'The life and work of E.G.R. Taylor, 1879–1966', *Journal of the Hakluyt Society* X, 1–12.

East, W. G. (1966), 'Professor A. C. O'Dell', *Geographical Journal* 132, 596.

Fairhurst, H. (1966), 'Alexander Stevens', *Scottish Geographical Magazine* 82, 58.

Freeman, T. W. (1980a), *A History of Modern British Geography*. Cambridge: Cambridge University Press.

Freeman, T. W. (1980b), 'The Royal Geographical Society and the development of geography', in E. H. Brown (ed.), *Geography, Yesterday and Tomorrow*. Oxford: Oxford University Press, 1–99.

Gilbert, E. W. (1966), 'Obituary: Professor J. F. Unstead', *Geographical Journal* 132, 334–5.

Maddrell, A. (2007), 'The "Map Girls". British women geographers' war work, shifting gender boundaries and reflections on the history of geography', *Transactions of the Institute of British Geographers* 33, 127–48.

Maddrell, A. (2009), *Complex Locations. Women's Geographical Work in the UK 1850–1970*. Chichester: Wiley-Blackwell.

Mead, W. R. (1995), 'Eila Muriel Joice Campbell, 1915–1994', *Imago Mundi* 47, 6–12.

Miller, A. A. (1943), 'Dr. H. A. Matthews', *Geography* 28, 63.

Myres, J. L. (1932), 'What is historical geography?', *Geography* 17, 39–45.

O'Dell, A. C. (1933), 'The urbanization of the Shetland Islands', *Geographical Journal* 81, 501–18.

Prince, H. C. (1995), 'Eila M. J. Campbell, 1915–1994', *Journal of Historical Geography* 21, 202–4.

Rawstron, E. M. (1985), 'Arthur Eltringham Smailes, 1911–1984', *Transactions of the Institute of British Geographers* 10, 120–2.

Richey, M. (2000), 'E.G.R. Taylor and the Vinland Map', *Journal of Navigation*, 193–205.

Robertson, C. J. (1973), 'Scottish geographers: the first hundred years', *Scottish Geographical Magazine* 89, 5–18.

Sheppard, J. (1994), *A Century of Geography*. London: Queen Mary College.

Skelton, R. A. (1951), 'Watermarks and the study of early maps', *Geographical Journal* 117, 221–2.

Skelton, R. A. (1968), 'Obituary: Professor E. G. R. Taylor', *Imago Mundi* 22, 114–16.

Spath, F. E. (1982), 'L.F. Spath, 1882–1957, ammonitologist', *Archives of Natural History* 11, 103–5.

Steel, R. W. (1984a), *The Institute of British Geographers: The First Fifty Years*. London: Institute of British Geographers.

Steel, R. W. (1984b), 'Harry Cyril Knapp Henderson, 1903–1983', *Transactions of the Institute of British Geographers* 9, 255–6.

Wallis, H. (1994), 'Eila Muriel Joice Campbell, 1915–1994', *Geographical Journal* 160, 361.

Willatts, E. C. (1971), 'Planning and geography in the last three decades', *Geographical Journal* 137, 311–30.

Willatts, E. C. (1987), 'Geographers and their involvement in planning', in R. W. Steel (ed.), *British Geography 1918–1945*. Cambridge: Cambridge University Press, 100–15.

Wise, M. J. (1975), 'A university teacher of geography', *Transactions of the Institute of British Geographers* 66, 1–16.

Worth, J. (2000), 'Christie Willatts', *The Guardian*, 17 March 2000.

2. SELECTED BIBLIOGRAPHY OF WORKS BY EVA TAYLOR

1910 (with J. F. Unstead) *General and Regional Geography for Students*. London: G. Philip (15th edn, 1952).

1911a (with J. F. Unstead) *Commercial Geography: General and Regional*. London: G. Philip (7th edn, 1933).

1911b (with J. F. Unstead) *Geography of the British Isles*. London: G. Philip.

1912a (with J. F. Unstead) 'Correspondence: Unstead and Taylor's "Commercial Geography"', *Geographical Journal* 39, 293–4.

1912b (with J. F. Unstead) *The Essentials of World Geography*. London: G. Philip (18th edn, 1954).

1915a *Questions and Exercises. Our Own Islands*. London: G. Philip.

1915b *Questions and Exercises. Lands Beyond the Channel.* London: G. Philip.

1915c *Questions and Exercises. Distant Lands.* London: G. Philip.

1916 'The first steps in geography teaching', *Geographical Teacher* 8, 224–30.

1919 *Long Ago and Far Away*. London: Constable.

1919 (with J. F. Unstead) *The British Lands and the Neighbouring Lands.* London: G. Philip.

1920 'Illuminating sketch maps', *Geographical Teacher* 10, 10–1.

1921 *A Sketch-Map Geography.* London: Methuen (Revised edn, 1950).

1922 *From Mountain to Sea.* London: Geographia.

1923 *The Business Man's Geography.* London: G. Philip (3rd edn, 1926).

1926–27 *A Complete World Geography*, 2 vols. London: Methuen (2nd edn, 1936).

1927 *Foundation Exercises in Geography.* London: G. Philip (13th edn, 1964).

1928 *Oceans and Rivers.* London: Benn.

1930a *Production and Trade: a geographical survey of all the countries of the world.* London: G. Philip.

1930b *Tudor Geography, 1485–1583.* London: Methuen.

1932a (ed.) *A Brief Summe of Geographie by Roger Barlow.* London: Hakluyt Society.

1932b 'Europe Centrale: a review', *Geographical Journal* 80, 60–2.

1933 'The cultural landscape: a review', *Geographical Journal* 81, 352–5.

1934a *Late Tudor and Early Stuart Geography, 1583–1650.* London: Methuen.

1934b 'The Geographical Congress at Warszawa', *Geographical Journal* 84, 427–9.

1934c 'The making of geography: a review', *Geographical Journal* 83, 66–7.

1935 (ed.) *The Original Writings and Correspondence of the Two Richard Hakluyts*, 2 vols. London: Hakluyt Society.

1936a 'Leland's England', in H. C. Darby (ed.), *An Historical Geography of England before A.D. 1800.* Cambridge: Cambridge University Press, 330–53.

1936b 'Camden's England', in H. C. Darby (ed.), *An Historical Geography of England before A.D. 1800.* Cambridge: Cambridge University Press, 354–86.

1938 'The geographical distribution of industry', *Geographical Journal* 92, 22–39.

1939 'A national atlas for Britain', *Nature* 144, 929–30.

1940a 'Plans for a national atlas', *Geographical Journal* 95, 96–108.

1940b 'A national atlas for Britain', *Nature* 145, 487–8.

1941 'A national master plan for Britain. Some fundamental facts', *The Architect*, 167–8.

1942 'Geographical aspects of regional planning', *Geographical Journal* 99, 61–80.

1942 (with G. J. G. B. Balfour) *Ground Plan of Britain: Fourteen Maps for the Use of Planning Activities and Groups.* London: The 1940 Council.

1943 *Land and Plan: Basic Facts Relative to a Master Plan for Britain.* London: Architect and Building News.

1945a *Planning Prospect.* London: Lund Humphries.

1945b *Geography of an Air Age.* London: Royal Institute of International Affairs.

1947a 'Richard Hakluyt', *Geographical Journal* 109, 165–74.

1947b 'Geography in war and peace', *Advancement of Science* 51, 187–94 (republished in *Scottish Geographical Magazine* 63, 97–108 [1947]; and in *Geographical Review* 38, 132–41 [1948]).

1949a 'The geographical basis of a county plan', *Journal of the Town Planning Institute* 35, 49–52.

1949b 'Urban geography: a review', *Geographical Journal* 114, 100–1.

1949c 'Road to survival: a review', *Geographical Journal* 113, 93–4.

1951 (ed.) *An Atlas of Tudor England, Being Forty Plates from John Speed's Pocket Atlas of 1627.* Harmondsworth: Penguin.

1952 'Geography of hunger: a review', *Geographical Journal* 118, 348–50.

1954 *The Mathematical Practitioners of Tudor and Stuart England.* Cambridge: Cambridge University Press (Reprinted in 1967 and 1970)

1956 *The Haven-Finding Art.* London: Hollis and Carter (New edition 1971).

1957 (with C. Singer and D. J. Price) 'Cartography, survey and navigation', in C. J. Singer and E. J. Holmyard (eds), *A History of Technology*, vol. III. Oxford: Oxford University Press, 501–37.

1959 (ed.) *The Troublesome Voyage of Captain Edward Fenton, 1582–1583.* Cambridge: Cambridge University Press for the Hakluyt Society.

1960 'Mathematics and the navigator in the thirteenth century' (Duke of Edinburgh's Lecture, 1959), *Journal of the Institute of Navigation* 13, 1–12.

1962 (with E. M. J. Campbell) *A Sketch-Map Geography.* London: Methuen.

1962 (with M. W. Richey) *The Geometrical Seaman. A Book of Early Nautical Instruments.* London: Hollis and Carter.

1963a (ed.) *A Regiment of the Sea and Other Writings on Navigation by William Bourne.* Cambridge: Cambridge University Press.

1963b 'The western invasion of the Pacific and its continents: a review', *Geographical Journal* 129, 536–7.

1966 *The Mathematical Practitioners of Hanoverian England, 1714–1840.* Cambridge: Cambridge University Press.

1974 'The Vinland Map', *Journal of Navigation* 27, 195–205.

3. SELECTED REVIEWS OF THE WORKS OF EVA TAYLOR

A.D. (1933), 'Late Tudor and Early Stuart Geography: a review', *Geography* 18, 226.

Anon. (1912), 'Two text-books of commercial geography', *Geographical Journal* 39, 151–2.

Baker, J. N. L. (1934), 'Late Tudor and Early Stuart Geography: a review', *Geographical Journal* 84, 171–2.

Calvert, H. R. (1966), 'The mathematical practitioners of Hanoverian England: a review', *Geographical Journal* 132, 520–1.

Crone, G. R. (1931), 'Tudor Geography: a review', *Geographical Journal* 77, 289–90.

Crone, G. R. (1955), 'The mathematical practitioners of Tudor and Stuart England: a review', *Geographical Journal* 121, 104–5.

Darius, J. (1990), 'The mathematical practitioners of Hanoverian England: a review', *Bulletin of the Scientific Instrument Society* 27, 31–2.

Johnson, F. R. (1935), 'Tudor Geography, and Late Tudor and Stuart Geography: a review', *Isis* 23, 289–94.

Johnson, F. R. (1957), 'The mathematical practitioners of Tudor and Stuart England: a review', *Isis* 48, 377–8.

Naish, G. P. B. (1957), 'The haven-finding art: a review', *Geographical Journal* 123, 85–6.

Newbigin, M. I. (1929), 'Oceans and rivers: a review', *Geographical Journal* 73, 180–1.

Parks, G. B. (1931), 'New light on Tudor Geography', *Geographical Review* 21, 686–7.

Parks, G. B. (1934), 'Sequel to Tudor Geography', *Geographical Review* 24, 693–4.

Parks, G. B. (1936), 'The Hakluyt writings', *Geographical Review* 26, 527–8.

Price, D. J. (1955), 'The mathematical practitioners', *Journal of the Institute of Navigation* 8, 12–16.

Scott, J. F. (1967), 'The mathematical practitioners of Hanoverian England: a review', *British Journal for the History of Science* 3, 300.

S. E. J. B. (1931), 'Production and trade: a review', *Geography* 16, 322.

Stamp, L. D. (1931), 'Production and trade: a review', *Geographical Journal* 77, 486.

W. E. W. (1929–30), 'Foundation exercises in geography: a review', *Geography* 15, 63.

Williamson, J. W. (1960), 'The troublesome voyage of Captain Edward Fenton, 1582–1583: a review', *Geographical Journal* 126, 236.

Chronology

1879	Born 22 June in Highgate, north London
1891–9	Secondary education at Camden School, and North London Collegiate School
1899–1903	Studied chemistry at Royal Holloway College

1903	Awarded BSc, first class honours, University of London
1905–6	Taught in Burton-upon-Trent and in Oxford
1906–8	Studied at the School of Geography, University of Oxford
1908–10	Served as research assistant to A. J. Herbertson
1910–16	Wrote textbooks and started a family; Spencer born 1912
1916–20	Taught part-time at colleges in London; Anthony born 1919
1921	Appointed to a part-time lectureship at Birkbeck College
1929	Awarded DSc, University of London for 'Studies in Tudor Geography'
1930	Publication of *Tudor Geography*; made Professor of Geography in the University of London, tenable at Birkbeck College
1934	Publication of *Late Tudor and Early Stuart Geography*
1939	Elected president of section E of the British Association for the Advancement of Science; re-elected 1947
1944	Retired from Birkbeck College, made professor emerita
1947	Awarded Victoria Medal of the Royal Geographical Society
1948	Awarded Honorary LLD, University of Aberdeen
1954	Publication of *The Mathematical Practitioners of Tudor and Stuart England*
1956	Publication of *The Haven-Finding Art*
1959	Delivered the first Duke of Edinburgh Lecture, Institute of Navigation
1960	Made Fellow of Birkbeck College
1965	Made Honorary Fellow of the Royal Geographical Society
1966	Publication of *The Mathematical Practitioners of Hanoverian England*; dies 5 July, in Wokingham, Berkshire

Orlando Ribeiro

1911–1997

Suzanne Daveau

Education, Life and Work

CHILDHOOD AND YOUTH

Orlando Ribeiro was born on 16 February 1911 into a family of small shopkeepers and craftsmen from northern Portugal. His paternal grandfather was a typographer in Viseu and his father worked as an assistant in a hardware shop in Porto for 13 years. His maternal grandfather from Chaves joined the army very young as a sergeant and retired a major. In 1909, his father, newly married to a cousin also from Viseu, settled in Lisbon where he had bought a share in a hardware shop near the Science Faculty, Botanic Gardens and Government Stationery Office. Orlando was placed in the care of his maternal grandparents as his mother's health had deteriorated after the birth of his brother Fernando. During World War I, he lived with them in the Home for Discharged Soldiers in Runa, near Torres Vedras, where his grandfather was an administrator. His mother died in 1916 and three years later he came back to live with his father, brother and stepmother, although he still spent his holidays with his grandparents who had retired to Viseu.

Ribeiro attended the Passos Manuel High School for his seven years of secondary education. In 1928, he went to the Arts Faculty of the University of Lisbon to read for a degree in History and Geography. An excellent student overall, despite living in Lisbon with a family with no intellectual interests, he started reading in the school library and the public library in a nearby park, borrowed books from his better-off fellow students, or read during his holidays in Viseu. There thus arose within him a vocation as a historian initially, then, more, as a geographer, when he read Emmanuel de Martonne's *Le Traité de Geographie Physique,* followed by *Les Principes de Geographie Humaine* by Vidal de La Blache (On de Martonne and Vidal de la Blache, see *Geographers* Vol. 12). In 1927, he attended classes in elementary Arabic taught by David Lopes as he wished to prepare himself to become a medieval historian, 'better even than Alexandre Herculano'. Ribeiro later said that the Arts Faculty was an average school, with a few excellent teachers. In his *Memoirs* and in festschrift articles, he mainly recalled two of them, who unfortunately died shortly after he started university: Manuel Ramos, a historian who became blind and had to be helped to prepare his lectures by his students, whom he rewarded generously for their assistance by following their education closely and taking them to concerts; and Silva Telles, a doctor from Goa, who established higher education in Geography in Lisbon in 1904. He had a sharp mind and was an excellent teacher, but he had done no in-depth personal fieldwork research.

At that time, the only true Portuguese geographer was Amorim Girão, professor at the University of Coimbra. In 1922, he was the first Portuguese person to submit a doctoral thesis in geography, on the Vouga Basin. He was virtually self-taught, claiming that his only influence was the geologist, Ferraz de Carvalho, and he never travelled outside Portugal. Girão left a significant, original and diverse body of work, however, including a proposal for reforming the administrative divisions of the country (1933). He also produced two works aimed at a general readership, a *Geography of Portugal* and an *Atlas of Portugal*, both published in 1941. Only one of his successors at the University of Coimbra, José Pereira de Oliveira, acknowledged the debt that his training owes to Girão.

In 1930, Orlando Ribeiro began to associate with José Leite de Vasconcelos, an elderly scholar who had already retired by that time from the Arts Faculty in Lisbon. Leite de Vasconcelos had an aristocratic provincial background and medical training, and had a keen interest in the ethnology, archaeology, numismatics and linguistics of the Portuguese people. He completed his training in France and Germany and created the Portuguese Ethnographic Museum (*Museu Etnográfico Português,* now called the *Museu Arquelógico Nacional*) in 1893 in Belém, in an annex of the Monastery of the Hieronymites. His correspondence with Orlando Ribeiro from 1931 and 1941, recently published in 2011, casts useful light on those crucial years in the education of the young geographer.

While earning his living by working as a teacher at the Colégio Infante de Sagres, a renowned secondary school, Orlando Ribeiro strove between 1932 and 1936 to broaden and round off his education, while writing his doctoral thesis on the Arrábida region, a small coastal mountain range on the left bank of the River Tagus, south of Lisbon. He attended the practical classes and lectures by the Swiss geologist, Ernest Fleury, who was teaching at the Higher Technical Institute (*Instituto Superior Técnico*). This provided Ribeiro's first introduction to fieldwork. He also learned how to use documentary records by helping Leite de Vasconcelos at home and obtained a small grant from the National Education Committee (*Junta da Educação Nacional*), an organization which, later, would become the IAC or Higher Cultural Institute (*Instituto de Alta Cultura*). This grant enabled him to travel throughout Portugal. His studies during the summer of 1936 focused in particular on the north-eastern part of the country called Trás-os-Montes, and he travelled in the company of Leite de Vasconcelos. In August 1935, Ribeiro had the opportunity to go on an Atlantic cruise organized by the Portuguese government to introduce young university students to the vast colonial Portuguese empire. He was thus able to call in succession at the port of Bissau in Portuguese Guinea, at the islands of São Vicente and Santiago in the Cape Verde archipelago and at various ports in Angola, from Cabinda to Moçãmedes, as well as Funchal in Madeira on the return journey. He was also able to spend several days driving across the huge inland plateaus of Angola. A young law professor, Marcelo Caetano, was responsible for the scientific aspect of the cruise: after holding several important posts in political and academic circles, he was eventually to replace Salazar at the head of the Portuguese government in 1968. Two young geographers, barely out of university, Orlando Ribeiro and Norberto Cardigos, acted as his assistants.

PARIS AND THE WAR YEARS

After defending his doctoral thesis on Arrábida on 28 March 1936, the young Ribeiro was nominated by Marcelo Caetano and A. Celestino da Costa, a medical doctor who was President of the *Junta da Educacão Nacional,* for a position as a Portuguese language assistant at the Sorbonne in Paris. He held this position from 1937 to June 1940. As a language assistant and an assistant to the Professor of Portuguese Georges Le Gentil, Ribeiro had to perfect his knowledge of Portuguese literature and history. His responsibilities ranged from teaching the basics of the Portuguese language to supervising doctoral theses. He took advantage of his stay in Paris mainly to complete his training as a geographer. In particular, he became a follower of Professor Emmanuel de Martonne, the Director of the *Institut de Géographie* and a world expert in geomorphology, and of Albert Demangeon (*Geographers* Vol. 11), who focused on issues relating to human and economic geography. Every summer, he would return to Portugal for a long spell with the family he started when he married the daughter of his former professor, Manuel Ramos, in 1936. He took advantage of this time to carry out fieldwork. With his two Parisian mentors, he decided to prepare a French doctoral thesis on the central region of Portugal, a crossroads where northern and southern influences met and mingled on physical and human levels. His Parisian contacts were not restricted to the geographers who taught at the Institut de Géographie in the rue Saint Jacques, attached to the Arts Faculty. He also attended lectures at the Science Faculty given by the geologists Luteaud and Bourcart, and at the Collège de France by André Siegfried (*Geographers* Vol. 30), Lucien Febvre and others, never missing an opportunity to broaden his knowledge and take part in a varied and intense cultural life.

Ribeiro's stay in Paris provided an opportunity to establish contacts with many geographers of his generation. French geographers, of course, some of whom became very dear friends, such as Pierre Birot, Jean Gottmann (*Geographers* Vol. 25) or Jean Demangeon, but also students from various countries who had come, like him, to complete their training in what was then the main and most active centre for geographical research: these included Frans Dussart, Ronald Harrison Church and Hassan Awad. Annual 'inter-university' field trips enabled him not only to discover the diversity of French regions, but also to get to know academics from other universities such as Henri Baulig, Roger Dion, Daniel Faucher and Jules Sion (see *Geographers* Vols 4, 18, this volume and 12, respectively).

In 1938, after a period of study in Belgium, Orlando Ribeiro attended the 15th International Geographical Congress in Amsterdam, where he read a paper on 'Rural settlement in Portugal'. This gave him the opportunity to meet most of the international geographical elite, in particular the German professor Hermann Lautensach, who presented his famous methodological approach to regional geography, and with whose important works on the geography of Portugal he already was familiar, courtesy of Leite de Vasconcelos and Amorim Girão. At the end of the Conference, the young Ribeiro, encouraged by de Martonne, was able to announce that Portugal's government was putting itself forward as a candidate to host the next International Geographical Congress, in 1942. But World War II broke out in 1939 and there then followed a long period of uncertainty and tension during which Portugal found itself more and more physically isolated within Europe. Since the summer of 1936, contact between France and Portugal had been restricted to sea travel, with the occasional stopover in an English port. Although he had been advised against it by his Parisian mentors, Orlando Ribeiro returned to Paris in 1940 to take up the post of Assistant Lecturer in Geography which Pierre Birot, who had been called up, could not therefore fill. Ribeiro shared this position with an Egyptian, Hassan Awad, and with Jean Gottman, who had just taken French citizenship. He left Paris by train on 7 June, barely a week before the

German army entered the city. He did not find a stable job in Portugal until February 1941 as a junior lecturer (*Professor Extraordánirio*) in Geography in the Arts Faculty of the University of Coimbra, where he taught for two years, before being appointed to the Chair of Geography (*Professor Catedrático*) in the Arts Faculty of the University of Lisbon. He held this Chair for 38 years, from March 1943 until retirement in February 1981.

Ribeiro's extensive correspondence, most of which has been preserved, is currently being catalogued. Study of it demonstrates that he did not lose touch with friends and colleagues whom he met in Paris just before the war, and that his privileged position in one of the few countries in Europe which remained neutral enabled him to act as a 'mailbox' to help those who could no longer write to each other directly. He also offered accommodation to scientists of various political persuasions fleeing the war and racial persecution, or trying to reach England, North Africa or America via Lisbon. Young Portuguese people who had suddenly returned to their homeland after studying abroad, were more than willing to welcome these foreign specialists, who gave them support and supervised their studies. Between 1940 and 1945, geological, geomorphological, prehistoric, sociological and ethnological studies flourished in Portugal thanks to the combined efforts not only of Orlando Ribeiro and the geologist Carlos Teixeira, both returned from France, but also from the work of the engineer and palaeontologist Mariano Feio and the anthropologist Jorge Dias, as well as the Franco-Russian, Georges Zbyszewski, who had settled in Portugal, and Henri Breuil, a priest, who would spend almost two years there as a political refugee before continuing his prehistory studies in South Africa. Jean Gottmann's stay in Lisbon as he fled from the south of France to America in 1941 was shorter. Pierre Birot, repatriated on medical grounds after three years in a prison camp in Germany where he helped organize a virtual university annexe, did not return to Portugal until 1943 to continue the geomorphological research he had begun in 1938.

In Lisbon, then a hub for intrigue and underground scheming between the warring nations, Ribeiro also welcomed (in 1943) the French geographer Pierre Deffontaines (*Geographers* Vol. 30), who had just been dismissed from his post as Director of the French Institute in Barcelona by the Vichy government. Hermann Lautensach, professor at the University of Greifswald, was allowed by the German government to spend two successive periods of research in Portugal. Accompanied by Ribeiro, Lautensach would go on long field excursions from Trás-o-Montes to Coimbra via the central mountain range, then to the Alentejo region and the Algarve. Together they planned a geography of Portugal in Portuguese, based on earlier work by Lautensach but updated and supplemented. In 1944, Emmanuel de Martonne managed to spend a few weeks in Portugal in order to familiarize himself with the field in which his two followers, Pierre Birot and Ribeiro, were working.

Franco's victory brought an end to the Spanish Civil War in March 1939. From 1940, Ribeiro finally succeeded in establishing contact with geographers in Spain via the Luso-Spanish Conferences for the Advancement of Science held in Saragossa in December 1940, in Porto in 1942 and in Cordoba in 1944 (Capel 1975). But his contacts developed mainly from 1946 onwards during research trips to Spanish Morocco and to Andalucia, which, he later recounted, were as important in terms of discovery as the Atlantic cruise of 1935. In July and August of 1946, he took part in the *Curso de Geografica General y del Pirineu* held in the small town of Jaca in the foothills of the Pyrenees. He there forged long-lasting links with two great geologists, the father and son Eduardo and Francisco Hernandez Pacheco, and geographers such as Manuel de Terán, Cassas Torres, Solé Sabaris and Adela Gil Crespo, with whom he travelled across the Sierra de Gredos in 1947. Ribeiro became increasingly knowledgeable about Spain through this extensive travel and, from 1979, via periodic meetings between geographers from both countries organized by the Iberian Geographical Colloquia (*Colóquios Ibéricos de Geografia*).

The end of World War II marked an important turning point in working conditions for Portuguese researchers, in both the social and the natural sciences. Relations between the outside world and Portugal were restored. The Atlantic sea routes were no longer perilous and new air links were created to destinations in Africa. Having cautiously kept a neutral position during the war, Salazar was led to believe that he could maintain Portugal's traditional peaceful position in this respect. Some of the technocrats around him, however, saw Portugal's problems in quite different ways, namely that of a former colonial empire, now reduced to a few African possessions bordering the Atlantic which Ribeiro had seen briefly in 1935, but whose main constituent country, Angola – which had an area of some 1,250 000 square km, salubrious highlands and mineral resources – appeared to be ripe for exploitation. In 1945, the *Junta das Missões Geográficas e de Investigações Coloniais* (Committee for Geographical Missions ['cartography' in a strict sense of the term] and Colonial Research) which had been created in 1936, was restructured and revived. In April 1946, Orlando Ribeiro was appointed a member and suggested a first academic geographical study mission to the Cape Verde Islands. That project was replaced by a reconnaissance trip to Portuguese Guinea, from January to May 1947. This had been deemed urgent because of the imminent Second West African Conference due to take place in Bissau in December 1947 (the first conference had been held in Dakar in 1946 at the initiative of Theodore Monod). From this time, the vast field of tropical studies opened up for Ribeiro and for his colleagues, opportunities which they would explore intensively over the following decades.

THE FIFTEENTH INTERNATIONAL GEOGRAPHICAL CONGRESS

Just as sea and air routes were experiencing an upsurge in activity which made it possible to embark on the study of territories outside Europe, so too international relations between major so-called developed countries involved in scientific research again flourished. The plan for an International Geographical Congress was taken up again by the International Geographical Union whose restored executive committee met first in London in July 1946, then again in Paris in July and November 1947. Since Portugal had renewed the invitation made in 1938, it was confirmed that the next Congress would indeed take place in Lisbon in April 1949. Ribeiro was appointed as secretary of the Organizing Committee.

With hindsight, it is not obvious how this International Conference with 779 participants from 38 countries and 206 papers was organized by fewer than 10 Portuguese geographers, since only a few of whom were actively involved in preparing and running the Conference and its associated excursions. The sessions took place in the modern premises of the Advanced Technical Institute (*Instituto Superior Técnico*). Thanks to grants, in particular from the newly created UNESCO, Ribeiro was able to find various kinds of ad hoc help and to focus it within the Centre for Geographical Studies which he had created in 1943 with support from the IAC, the Higher Cultural Institute (*Instituto de Alta Cultura*), successor organization to the *Junta de Educação Nacional*. The problem was compounded by the fact of there being very few geography students. Only two female students were actively involved in the preparations for the Conference, Maria Plácido Santos, and Raquel Soeiro de Brito, who later pursued an academic career as a geographer. Ribeiro's broad vision of geography, which he viewed as a meeting point for many disciplines, allowed him to hire as his assistant Mariano Feio, an engineer who was to work for him for many years as a reliable and brilliant 'right-hand man' and who, in 1952, submitted a remarkable doctoral thesis on the geomorphological evolution of southern Portugal. Ribeiro also hired as a temporary secretary a young poet, Francisco Tenreiro, who had a background in sociology. He, in turn, submitted an excellent doctoral thesis in geography in 1961 focusing on his native island of São Tomé. With these as helpers, Ribeiro organized the Lisbon meeting.

The six main excursions which took place as part of the IGU Conference were each provided with a guide book in French, which formed the basic bibliography on Portugal for all non-Portuguese or non-German readers for decades thereafter. Ribeiro prepared and led two of the excursions, 'Central Portugal' and 'The Island of Madeira'. An assistant from Coimbra, Fernandes Martins, was responsible for 'The Coastal Central Region' and 'The Limestone range of Estramadura'. Mariano Feio wrote 'The Lower Alentejo and the Algarve'. The 'Estramadura and Ribatejo' excursion was presented by the geologist, Zsysewski, in collaboration with Virgínia Rau, a historian, while the 'Minho, Trás-os-Montes and Douro' was presented by the geologist, Carlos Teixeira, and the anthropologist, Jorge Dias.

1949 was not just 'the year of the Conference'. For Ribeiro, it was also the year when he first publicly expressed his views about the university system and scientific research. On 31 October, against a centuries-old background of rivalry with the University of Coimbra, the University of Lisbon decided to revive the obsolete old tradition of the Formal Opening of the Academic Year, during which an Address on Wisdom or on Knowledge (*Oração de Sapiência*) was delivered. The responsibility fell to the youngest lecturer, Orlando Ribeiro. He took advantage of this to present a highly critical picture of a university system which he considered to be fossilized, and to suggest some necessary remedies. According to him, university teaching should be closely linked to research, so that a lecturer was required to be 'a researcher who teaches', then a great rarity. At the same time, the *Junta de Investigações Coloniais* asked him to take part in a series of 18 *colóquios,* designed 'to finalize and freely discuss' the studies then being promoted by this new organization. The other contributors spoke only about their own specific research topic, generally relating to the natural sciences, but Ribeiro chose to speak about a much more general theme, namely 'Problems arising from colonial scientific research'. After reminding them, respectfully, that, contrary to received patriotic ideas, tropical scientific research was not born in Portugal, he demanded the same freedom and quality for overseas work as he had two months earlier done for research relating to mainland Portugal.

The position which he would continue to occupy for a long time in Portuguese academic circles was already clearly apparent. He was a young researcher with a world-wide reputation (he was to be promoted first Vice-President of the IGU at the Washington Conference in 1952). But he was, by contrast, also considered at a national level to be an 'awkward customer' because he spoke his mind in public, while refusing to adopt a political stance, and because he made several attempts to push through his reform plans. At the Lisbon Conference, for example, he had already made clear for the first time the strength of his commitment to international research and scholarship by persuading the Portuguese government to invite Hermann Lautensach at a time when German geographers were excluded from the IGU, as they had been after World War I.

THE CREATION OF A SCHOOL OF GEOGRAPHY

The years after 1945 were very active for Ribeiro. He received frequent invitations as a result of his international contacts and these provided an opportunity to visit various countries and to take part in meetings and scientific excursions. He in turn welcomed foreign colleagues to Portugal, with or without their students. He would almost always go and work in the field with his visitors. When the Conference fund was exhausted and its Proceedings circulated, the work of the Centre for Geographical Studies returned to normal, limited by a meagre budget which was boosted by occasional subsidies from the *Junta de Investigação Colonial* (which became the *Junta de Investigações do Ultramar* in 1951). But the fact that students were scarce gave researchers time to carry out fieldwork. Ribeiro welcomed young researchers specializing in various subjects to the Centre, where they would at least find a stimulating geographer to listen to. The Italian researcher, Gaetano Ferro, came to prepare a doctoral thesis on the Algarve (submitted in 1956); the French historian, Albert Silbert, was welcomed there as he wrote his monumental thesis on agrarian history relating to Mediterranean Portugal at the end of the *Ancien Régime* (submitted in 1966); and the Latin scholar and ethnologist Manuel Viegas Guerreiro was hired in 1954 as a member of staff at the Centre to catalogue and publish the manuscripts of Leite de Vasconcelos. Sponsored by Ribeiro, Guerreiro was responsible for the publication between 1958 and 1988 of seven of the ten volumes of the monumental *Etnografia Portuguesa,* prepared by Leite de Vasoncelos.

Although the Centre for Geographical Studies was managing to develop gradually, despite the lack of any regular and significant financial support, the training of young geographers remained inadequate. From 1930, academic geography, which had been viewed until this point as a 'subsidiary science', at last broke free from history. But geography had not yet achieved sufficient cohesion to attract many students, especially since a degree in 'geographical sciences' offered no direct opening into secondary teaching. After spending two years in the Science Faculty devoted to various branches of the natural sciences and one certificate in Physical Geography (taught by a geologist), students transferred to the Arts Faculty where some teaching of Human Geography with a little history was provided. The course content focused in particular on colonial issues, and included certificates in the History of Discoveries, Ethnology and Colonial Geography. Very few students chose this course, sometimes no more than one student a year, two or three at the most. In 1968, Orlando Ribeiro called this a 'disastrous reform' which had created a long-term shortage of well-trained geographers in secondary and higher education. Not until 1957 did the reform of teaching in Arts Faculties at last include a better balanced Geography degree. By contrast with the previous system, it was a long course of five years' duration because it included an extra certificate in applied geography, two 'options' and a final year research project. This new degree attracted more students, however, at first mostly young women who wanted to be secondary school teachers. By 1960, the situation had improved significantly in Lisbon, with the joint arrival of three students, Jorge Gaspar, Carlos Alberto Medeiros and António de Brum Ferreira who became followers, even 'disciples', of Orlando Ribeiro and later successors to him. From that point, the training of young geographers became one of the priorities for the small number of geographers. In time, the gradual increase in numbers – because of the associated development of secondary education in Portugal – was such that large student numbers became as difficult to manage as the previous shortage.

In June 1951, a volcano erupted suddenly on the island of Fogo in the Cape Verde Islands, thus ending almost a century of inactivity. Ribeiro at once volunteered to study it, taking advantage of the opportunity to resume a previous research project in the archipelago. After a second stay there from December 1952 to January 1953, he published a monograph on the island as well as on all those eruptions which had taken place there since its discovery (Ribeiro 1954). In it he described the famines which used to occur, regularly until 1947–8, when the summer rainy season failed, causing an appallingly high death rate among this island people. He suggested simple administrative and economic measures which should make it possible to avoid such disasters. This element of his 1954 book caused a scandal among the Portuguese authorities who preferred to draw a discreet veil over such unfortunate problems. Thanks, however, to the author's international reputation, the book was spared, although its distribution was delayed for nearly three years: it was reissued in 1960. As a result, the next rain-free spell, in

1959–60, no longer led to a rise in mortality. In 1964, Ilídio do Amarala, a follower of Orlando Ribeiro, submitted a doctoral thesis in the form of a detailed monograph of the Cape Verde island of Santiago.

In 1951, an invitation to a scientific conference in Cairo provided an opportunity to explore the Nile valley and to get to know Greece and Italy on the return journey. In the spring of 1952, Ribeiro took advantage of the maiden voyage of the Portuguese liner *Santa Maria* from Lisbon to Rio de Janeiro to 'discover America'. Having received a warm welcome in Brazil, he travelled across most of this huge country, paying his way with a series of invited lectures. He travelled from Santa Catarina in the south to Amapa and French Guyana, where he acted as an interpreter during an encounter near the border. He went to Peru, which he crossed by train from Lima to Cuzco and Machu Picchu, and eventually entered the United States, to participate in the International Geographical Congress in Washington in 1952, where he was elected First Vice-President of the IGU. He also travelled across the eastern United States, from New York to Alabama. In 1952, he also attended the Third West African Conference organized on the Spanish island of Fernando Pò, and he took advantage of that trip to study the neighbouring Portuguese islands of Príncipe and São Tomé. During the following years, the Azores archipelago was to claim his attention, first in relation to his supervision by Raquel Soeiro de Brito a doctoral thesis on the island of São Miguel (submitted in 1955), then, in 1957, to study a new volcanic eruption on the island of Fayal. This eruption was filmed by his colleague and observed by his eldest son, António Ribeiro, aged 12, for whom it may have marked the beginning of his future vocation as a geologist (Figure 1).

Ribeiro's main 'discovery' during the 1950s stemmed from a short but disturbing encounter with the Indian world. Tension was increasing between Salazar's Portugal, still clinging to the remains of a century-old colonial empire, and the young Indian Union, created in 1946. After the bloody failure of an Indian 'peaceful march' on Portuguese territories in August 1955, the Indian Union broke off diplomatic relations with Portugal. Ribeiro remained convinced of the depth of the century-old influence

Figure 1: The Capelhinos eruption on Faial Island, in the Azores, 1957. Photograph courtesy of the author.

of Portuguese civilization on this small corner of India, an influence to which the pages devoted to Goa by O. H. K. Spate in his major work *India and Pakistan* (1954) bear witness. The somewhat utopian 'Portuguese-tropicalist' interpretation of Portuguese humanist expansion based on interbreeding and integration had just been circulated, and the Brazilian sociologist Gilberto Freyre had recently extended it to India in a book published in 1953. Orlando Ribeiro felt that a serious scientific study of the country would perhaps find common ground with India, especially in the light of his trip across Brazil.

Deep disappointment awaited the three Portuguese geographers who landed in Goa in October 1955, after calling at Cairo and Karachi. They were confronted by a general lack of familiarity with the Portuguese language, a society which remained foreign and indifferent, if not hostile, a very limited European influence, or 'a cyst on the flank of Hinduism in full revival', as Ribeiro wrote. Upon return to Lisbon in April 1956, he decided to send his text directly to Salazar, who was President of the Council, well aware that it was bound to get lost as it went through the usual administrative channels. Although Salazar thanked him politely and mentioned it to the President of the Republic as being of interest, Ribeiro's report had no impact on government policy. The Portuguese territories were occupied by force in April 1962 and annexed by the Indian Union. Half a century later, in 1999, the report was published posthumously courtesy of the National Commission for the Commemoration of Portuguese discoveries (*Comissão Nacional para a Commemoração dos Descobrimentos Portugueses*).

This disappointment did not deter the geographer from remaining active. In 1955, a *Geography of Portugal* was published in Spanish in Barcelona, the fifth volume of the *Geografía de España y Portugal,* edited by Manuel de Terán. In 1956, Orlando Ribeiro took an active part in the 18th International Geographical Congress in Rio de Janeiro and the associated Course in Higher Geographical Studies (*Curso de Altos Estudos Geográphicos*). Under the supervision of Hilgard O'Reilly Sternberg, the course was delivered by two French geographers, Pierre Birot and Pierre Deffontaines, the geologist André Cailleux, by the American geographer and cartographer Erwin Raisz (*Geographers* Vol. 6), the German geographer Carl Troll (*Geographers* Vol. 3) and by Ribeiro, whose chosen topic was 'The Mediterranean and the Organisation of the Tropical World'.

In January 1959, Ribeiro gave five lectures at the Collège de France on 'The geography of Portuguese expansion'. That year was marked by the failure of his attempt to reform the Arts Faculty of the University of Lisbon, an institution at which he had studied and where he had taught since 1943. Thanks to the support promised to him by the Principal of the University, Marcelo Caetano, as well as the Minister for Education, Leito Pinto, he accepted the appointment of Head of the Faculty and he rapidly handed the authorities a proposal for reforms designed to shake up the traditional lethargy of the Faculty, based on a close relationship between scientific research and teaching, as he had already outlined in 1949. He also wanted to ensure that the youngest teaching staff had better working and career conditions, allowing them to devote themselves purely to academic life, and to revive the Alumni Association, as he had attempted to do in the past when he was a student. The Principal and the Minister backed away from these reforms given the hostile reception they had received in high places and advised him to be patient. Since patience was hardly his strong point, Ribeiro resigned at the end of September upon his return from fieldwork in Trás-os-Montes, France and Spain. This bitter episode confirmed his ambiguous position within academic and political circles in Portugal. As a respected scientist, protected by his international reputation, Orlando Ribeiro was, more than ever, considered an awkward individual in his own country, yet someone who could hardly be forced into exile (a fate to which many other colleagues had had to resign themselves), but who could be hindered and targeted indirectly at any time principally by attacking his more vulnerable colleagues.

In 1959, the Centre for Geographical Studies moved into the new premises which it still occupies within the new Arts Faculty on the University Campus on the northern outskirts of Lisbon. Although it enjoyed much better facilities, the Centre became a kind of annexe to a purely literary school, cut off therefore from its traditional contacts with the Science Faculty which remained in its old premises in the centre of Lisbon. Even so, the Centre for Geographical Studies was gradually developing by recruiting a series of young researchers writing theses in various fields after spending one or two years studying abroad: Ilídio do Amaral and Maria João Alcoforado in Germany; Jorge Gaspar in Sweden; Brum Ferreira, Carlos Alberto Medeiros, Carminda Cavacao and Isabel Medeiros in various French universities; Paula Bordalo Lema in England; Teresa Barata Salgueiro in the United States. A Sedimentology Laboratory was set up in the Centre, directed by the geologist Galopim de Carvalho, to strengthen research in Geomorphology. I married Orlando Ribeiro in 1965 and settled in Portugal to play an active part, in research initially, and then in teaching. I developed in particular the use of then new techniques such as remote sensing and interpretative cartography, using and updating a number of my husband's current works which his increasingly demanding and numerous responsibilities and already frail health had not allowed him to complete.

RESEARCH IN OVERSEAS TERRITORIES

1960 marked the start of an important new area of research for Portuguese geographers. It also hailed the beginning of a series of detailed research trips to the two large territories of Angola and Mozambique. Before describing them, it is important to recount an incident that occurred in Stockholm on the occasion of the plenary session of the 19th International Geographical Congress in Norden. When Professor Carl Troll was there officially proposed as the new President of the IGU, the announcement met with strong opposition from representatives from the USSR and countries with similar political tendencies, under the pretext that an atlas, published under the direction of Troll, showed a border with Germany which had been drawn in a way which would now be termed 'politically incorrect'. I was then astonished to see Orlando Ribeiro, whom I had just met during a trip in central Sweden, requesting permission to speak and calmly addressing a huge frenzied audience to remind them that the IGU was a purely scientific and hence apolitical organization. The choice of a President should take into account only the candidate's merit as a scientist. This brief but courageous speech was sufficient to ensure that Carl Troll was elected without difficulty. I then understood for the first time that my then friendly companion during the trip (and, later, husband) was brave enough to face on his own a crowd unsettled by the anguish of the Cold War and to do so on purely ethical grounds.

Leaving Stockholm on 12 August, Orlando Ribeiro arrived four days later in Lourenço Marques (now Maputo), the capital of Mozambique. He gave a lecture there which caused something of a scandal, on the human problems of Africa in which he demonstrated his view that concepts of races and the increased status given to them were unscientific. He thought at that time that immigration from mainland Portugal could be beneficial for Portugal as well as African countries, but on the condition that 'the various races could live in peace with dignity and respect for one another, based on understanding, tolerance and equality'. Although these simple yet daring words filled the audience with enthusiasm, they worried some of those responsible for a country whose economy was closely dependent on the South African Union. When the same lecture was repeated a few weeks later in Luanda (Angola), it received a mixed reception. It was swiftly published by the *Junta de Investigações do Ultramar*, and Ribeiro acknowledged that there were never any official repercussions.

During the following decade, before the colonial war broke out, he and his young colleagues were able to work totally freely in Africa and in much better conditions than in mainland Portugal. Thus, Fernandes Martins, a professor at the University of Coimbra, began research on the inselbergs of the Nampula region in Mozambique, while two other geographers began studying the relief of Angola, Mariano Feio in the south, Ilídio do Amaral in the north. The latter had temporarily to suspend his research when serious disturbances broke out in 1961 in the coffee-producing region. When colonial order was rapidly restored, however, the two Portuguese territories enjoyed a sudden and belated economic development in subsequent years. This meant that the geographers were able to move around freely. The present author was a professor at that time at the University of Dakar in Senegal, which had been independent since 1958, and which had no diplomatic relations with Salazar's Portugal and so, in September 1961, was able without any difficulty to study northern Mozambique and southern Angola with Ribeiro. Ribeiro arranged an invitation to Angola in 1962 and Mozambique in 1963 for his colleague and friend, the great tropical geographer, Pierre Gourou (*Geographers* Vol. 25).

During the final years of the colonial era, Salazar's New State (*Estado Novo*) undoubtedly had a more restrictive attitude towards science at home than it did in its distant African possessions, where decisions rested with younger men with a broader outlook. It was decided, albeit belatedly, to establish a higher education system in the two large African colonies. The followers of Ribeiro took up the first teaching posts in geography from 1970 in Sá da Bandeira, a remote and small colonial town on the southern plateau of Angola, and in Lourenço Marques in Mozambique. These young geographers devoted themselves enthusiastically to research and although they scarcely had the time or conditions required to train local followers due to insufficiencies in the facilities for basic teaching, they started a new phase in geographical knowledge in that area of Africa (Figure 2).

Figure 2: Orlando Ribeiro playing the balafon, Angola, 1963. Photograph courtesy of the author.

Despite the interruption caused by colonial wars, a series of important studies were published over the following decades. A few noteworthy examples include the study of *Luanda* by Ilídio do Amaral, one of *Maputo* by Clara Mendes, on *A Colonização das Terras Altas da Huíla* (The Colonisation of the Highlands of Huíla) by Carlos Alberto Medeiros, and, in 1981, as a form of overall assessment, *Colonização de Angola e o seu Fracasso* (The Colonisation of Angola and its Failure) by Ribeiro himself, in which he offered a broad comparative panorama of various Portuguese settlements overseas (Ribeiro 1981). This work was supplemented in 1998 by *As Causas do Fracasso da Colonização Agricola de Angola* (Causes of the Failure of Agricultural Colonisation of Angola) by Mariano Feio.

THE CONTRIBUTION OF THE GULBENKIAN FOUNDATION

The establishment of the Calouste Gulbenkian Foundation in Lisbon in 1956 ushered in a new era in Portugal. This organization inherited the huge oil fortune of its founder, a fortune that its lawyer President, Azeredo Perdigão, managed without political interference and devoted mainly to cultural activities hitherto woefully neglected in Portugal. The Centre for Geographical Studies began to receive subsidies from the Foundation in 1961 and its Director had grounds for believing that he would at last be granted a stable and adequate budget. Ribeiro was among the small number of high-ranking intellectuals from a variety of academic backgrounds well-known for their particular independence of mind recruited by the Gulbenkian Foundation to draw up a development plan for scientific research in Portugal. These researchers suggested creating a Gulbenkian Institute for Science (*Instituto Gulbenkian de Ciência*), to manage several research Centres. Ribeiro suggested the creation of an Institute of Earth and Man (*Instituto da Terra e do Homem*). Unfortunately, and for various reasons ranging from fluctuations in the price of oil to conflicting ideas among the various members of the Council on the best way to develop scientific research in Portugal, these plans eventually resulted in only a few uncoordinated and short-lived achievements. In particular, Orlando Ribeiro never managed to link the Centre for Geographical Studies directly to the Gulbenkian Foundation to ensure a stable budget.

For over a decade, however, the Centre had at its disposal noticeably increased financial resources, enabling it to carry out worthwhile projects. In particular, one long-standing ambition of its director was enabled to take shape: the publication of a geographical journal of international standing. *Finisterra, Revista Portuguesa de Geografia* was launched in 1966: its regular bi-annual publication has been maintained since then and now achieves international circulation. Two collections were also established at that time: *Chorografia,* which took the form of lavishly illustrated booklets to introduce students and teachers to local and regional studies, and *Memórias*, which was designed principally to publish doctoral theses and other detailed studies. In total, ten issues of *Chorografia* were published before 1986 and fifteen *Memorías* before 1992 when the editorial policy was modified. The excellent publications produced intermittently since the fifties by the *Junta de Investigações do Ultramar* are also worthy of note.

The Centre also provided resources to facilitate geographical work in other ways: to rent vehicles, books and maps; to hire technical staff and to finance scientific meetings, including the First International Geographical Seminar (*Primeiro Seminário Internacional de Geografia*), which had 200 participants. At this event, two French geographers, Etienne Juillard and Jean Demangeot, the Canadian botanist Pierre Dansereau and the Brazilian geographer, Orlando Valverde, presented innovative ideas on approaches to the various branches of geography. In 1980, the Second Iberian Geography Symposium (*Segundo Colóquio Ibérico de Geografia*) was held in Lisbon, the first such symposium having taken place the previous year in Salamanca. These meetings, which still take place, foster useful links between the ever-increasing number of young geographers from neighbouring countries.

THE CLOSING YEARS OF RIBEIRO'S CAREER

Orlando Ribeiro was Director of the Centre for Geographical Studies until 1970, when he handed over this onerous responsibility to his assistant Ilídio Amaral, before the revolution of 25 April 1974 temporarily threw the old structures into turmoil. In 1995, another follower, Jorge Gaspar, took over as long-term Director of the Centre. In the meantime, there had been a series of disappointments as a succession of attempts at reform and development largely failed. An academic year spent in Paris in 1967–8 as a visiting Professor at the Sorbonne was frustrating. Ribeiro no longer found the atmosphere he had experienced there during his youth and the student riots of May 1968 filled him with dismay. He still continued to travel and write despite his failing health, although he was aware that he would not fulfil the many dreams of his youth.

In 1970, his hopes of achieving meaningful reform of the Portuguese University system were raised within the framework of a modernization programme which the new Minister for Education, Veíga Simão, was then striving to introduce. Ribeiro led a group of academics of all ages and ranks in preparing a plan for a revised structure for what they called the Faculty of Human Sciences (*Faculdade de Ciêcias Humanas*). Unfortunately, this plan did not materialize until the day before the revolution of 25 April 1974 – and so was, of course, set aside and forgotten as a product of the former regime. Nevertheless, Orlando Ribeiro carried on writing and attempting to publish. Figure 3 situates the phases of development of his main works in the context of his long career. In 1968, a work which had evolved over a long period *Mediterrâneo Ambiente e Tradição* was published.

Tradição (*The Mediterranean: Environment and Tradition*) was published (Ribeiro 1968). This work was translated and reprinted three times in Italian. This was followed by a university textbook, *La Zone Intertropicale Humide*, co-authored with the present author (Ribeiro and Daveau 1973). Following that period of brutal and chaotic decolonization linked to the 1974 revolution, Ribeiro wrote two more books about the problems of overseas territories marked by Portuguese influence: *Destinos de Ultramar* (*What Future for Overseas Territories?*), in 1976, then *A Colonização de Angola e o seu Fracasso* (*The Colonisation of Angola and its Failure*) in 1981.

The final years of his career were mainly devoted to collecting and revising his dispersed manuscripts and articles, some of which were published in 1970 in a volume of *Ensaios Geográficos,* others gathered together in the six volumes of the *Opúsculos Geográficos* published by the Gulbenkian Foundation between 1989 and 1995. I published a four-volume *Geografia de Portugal,* bringing together the translation of part of the volume published by Lautensach in 1932–7 with the original version published by Orlando Ribeiro in Spanish in 1955, and with necessary updates. An exhibition organized in Coimbra in 1994 at the initiative of Jorge Gaspar, gave the general public an opportunity to discover for the first time, if rather belatedly, that Ribeiro was also a very able photographer.

(The different boxes and identifying labels signal the phases of Ribeiro's life and work and, in being read chronologically from left to right, to development of his work: from the initial idea or announcement, preparation, printing, pre-edition or annexe, complete manuscript or duplicated version, first edition, other édition, illustrated edition, translation, to date of any corrected edition.) A: *Academic life*: Childhood, High School Education, University of Lisbon, University of Paris, University of Coimbra, University of Lisbon, former facilities, new facilities, Retirement. B: *Main scientific works*:

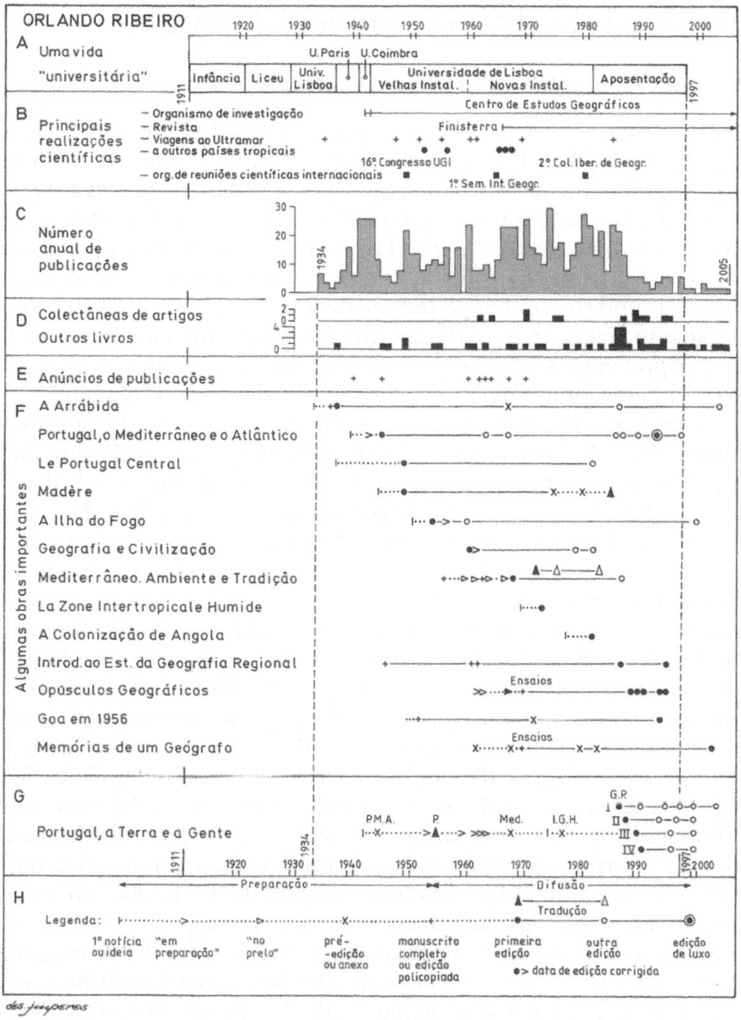

Figure 3: Planning and circulation of Orlando Ribeiro's main scientific works.

Research Organisation, Centre for Geographical Studies; the journal *Finisterra*; Trips to overseas territories; Trips abroad; Organisation of international scientific gatherings, 16th IGU Conference, 1st International Geography Seminar. C: Future publications D: Significant works E: Plans and versions of *A Geography of Portugal.*

In 1985, his last most enjoyable experience was to revisit the Cape Verde Islands at the invitation of the President of this newly independent state, in recognition of the help he had courageously given in the past to a population which still cherishes his memory. In March 1986, however, a brain haemorrhage suddenly put an end to the creative period of Orlando Ribeiro's life. He survived another 12 years, however, and remained interested in the publication of his own works as well as in those of his followers and friends. He left behind an important body of unpublished material which he had bequeathed to the Centre for Geographical Studies in Lisbon in 1983, together with his library and archives.

Scientific Ideas and Geographical Thought

RIBEIRO'S SCIENTIFIC AND CULTURAL APPROACH

Orlando Ribeiro summed up his views on geography in 1970 in that lucid, yet dense and poetic prose which lends his work such charm. It is, he recounted, 'both a basic science and a place where the sciences converge, a point of departure and a meeting place. It is like a crossroads which you can approach and leave by different routes'. He also specifies that his own progress was fuelled by two sources: 'firstly, a love of nature and the appeal exerted over me by solitude and peaceful mountains and deserted shores, dense forest shade and the contemplation of the solid architecture of the globe and its relief, the changing appearance of the sky and sea which surround it and all the things which our over-pretentious era claims would be the same with or without man'. He referred to the other inspiration for his calling as 'the slow tide of human history, which comprises neither dates nor figures, which is little known but has been transmitted by popular culture since the dawn of time to the present day, . . . following the rhythms of imperceptible changes merely hinted at by observation, but which the mind can reconstitute'. For him, geography was inseparable from history: 'It is the means of seeing and feeling what has survived for us from the past . . . Unless we carry out in-depth research into the past, our knowledge of most geographical facts remains incomplete and superficial' (Ribeiro 1970a, 17, 19).

Having aspired since his student days to devote his life to gaining an understanding of Portugal, Ribeiro reflected at several points on what geography should be. In 1934, he explained his interpretation of human geography. He defined it as 'the study of the relationships between human groups and the physical environment, a study based on their expression in aspects of the landscape' (Ribeiro 1934). Since he had no elder figure in Lisbon with whom to develop his ideas since Silva Telles died when he was in his first year at university, and the other professor of geography who would be his colleague from 1943 to 1955 had always struck him as intellectually limited, Ribeiro sought support from researchers in related sciences. These men included the geologist Ernest Fleury, who taught him how to observe terrain methodically, and, notably, Leite de Vasconcelos with whom he maintained a close relationship in the last 12 years of his life. Leite de Vasconcelos always demonstrated great respect for the young geographer's vocation and never tried to turn him into an ethnologist, but taught him, rather, how to carry out fieldwork, to catalogue methodically the documentation gathered and to write up the results obtained in the field in a clear and honest manner, with an unfailing enthusiasm for research.

In Lisbon and in Coimbra – the only two Portuguese cities at that time with Arts Faculties which taught geography – the narrow intellectual circles favoured contacts between disciplines. Maintaining their high-school friendships, students liked to attend lectures by renowned professors in other disciplines, or to go and listen to visiting artists. Members of the government at that time were nearly all university professors or, less commonly, architects or engineers. This was an extremely closed environment and it was not widely accepted (with the striking exception of Salazar) that locally acquired education should be rounded off by several years of study abroad. During World War II, the opposite became true: international scientific contacts were fostered in neutral, safe Portuguese territory, thus leading to the return of young bursary holders and a keen revival in the study of the natural and human features of the country.

Any attempt to define the geographical topics most explored and taught by Orlando Ribeiro encounters both great variety and should note the influence exerted by his own early education. In 1939 and 1940, he published on geomorphology, both on Central

Portugal and on the Serra de Sintra to the west of Lisbon. The influence of the teaching of Lautensach, and especially that of Emmanuel de Martonne, can be discerned in this work. Also evident is his daring handling of various theoretical explanations then in their infancy. Although he acknowledged William Morris Davis's stages of erosion of surfaces, Ribeiro was not afraid to suggest associated tectonic deformations and he used significant changes in climate to explain the arrangement of forms and deposits (on Davis, see *Geographers* Vol. 5). His writings on climate and vegetation in the regions studied may now seem less innovative, but they are still robust. Dynamic climatology was barely in its infancy at that time and geographers were largely unfamiliar with it, and palynologists were only just beginning to piece together the history of vegetation cover in northern Europe over recent millennia. Following the establishment of a Sedimentology Laboratory in the Centre for Geographical Studies in 1965, climate research in Portugal proliferated, followed later by studies of vegetation cover and the transformations effected by humans.

Orlando Ribeiro's initial human geographical research focused on the rural world, with the exception of an early article on the site and development of Lisbon, the city of his birth. In 1938, he circulated questionnaires relating to regional geography and rural settlement to try to assemble uniform documentation on a national Portuguese scale. This early research bears the hallmarks of the dual influence of Leite de Vasconcelos and Demangeon, but it is also possible to discern in it the early discovery of specifically Portuguese features both on a regional level, such as the seasonal migrations in the Serra da Peneda, the summer transhumance in the Serra da Estrela, and the survival of communal life in certain villages in Beira. On a national level, we can see here the establishment of a classification of the major types of animal husbandry or rural habitat, and the recognition of the diversity of regional development in population density depending on internal and external migration. Initially, the results were expressed in creative high-quality mapping, but this did not culminate in the planned Atlas of Portugal due to a lack of resources.

Over time, Ribeiro became increasingly interested in the historical and societal aspects of geography. This is evident in the titles of his best-known works: *Geografia e Civilização* (*Geography and Civilisation*, 1961) and *Mediterrâneo. Ambiente e Tradição* (*The Mediterranean: Environment and Tradition*, 1968). Analysis of the themes which provided the material for his teaching and the textbooks which he published reveals two publication phases in the latter context. From 1941 to 1943, he published a series of lessons for use in secondary education in the journal *Liceus de Portugal*. By contrast, material from the lectures which he delivered for decades in various universities was not published until much later. In 1973, he and I co-authored an academic textbook in French, *La Zone Intertropicale Humide* (*The Intertropical Humid Zone*) set in the context of major zones of civilization. Not until the later stages of his career did he publish those carefully thought-out lessons presented in a deliberately accessible fashion which probably best represent his 40 years as a professor: *Iniciação em Geografia Humana* (*Introduction to Human Geography*, 1986) and *Introdução ao Estudo da Geografia Regional* (*Introduction to the Study of Regional Geography*, 1987).

Rather than limiting his research to a few branches of geography, and while sending many of his followers to train abroad in various emerging disciplines which they had freely chosen, Ribeiro – effectively, the founder of modern Portuguese geography – always maintained the same position by combining study of the natural base with attention to the influence of human activities on the environment. He could be defined, fundamentally, as a humanist. This did not prevent him, however, from being open to new methodologies be they technical, such as sedimentology or remote sensing or conceptual, such as theoretical models for a so-called New Geography. He gave his followers free rein to experiment, while always keeping a close critical eye on their experiments. Upon Ribeiro's return to Portugal in 1940, he had been widely consulted as an expert on a wide variety of topics such as the construction of the new high school in Coimbra, the drafting of various urban development plans and the design for districts for members of the poorer social classes. Most of these experiences left a bitter taste in Ribeiro's mouth, as his opinion was either not always taken into account or, worse, it was completely distorted. In degree programmes he was more successful. He introduced a compulsory 'Applied Geography' topic into the degree course defined in 1957. He would always stress to his followers that any high-quality geographical study is both useful and has applications, but that work produced to order was rarely used by those who commissioned it when the conclusions appeared inconvenient. For him, the most important thing was to retain the option to circulate the results obtained freely. He always considered his greatest professional success to be putting an end to the abnormally high death rate in the Cape Verde Islands in rain-free years due to the moral courage he displayed in eventually getting his book published in the face of official dissent.

Ribeiro always emphasized the need for detailed and extensive fieldwork and reiterated to students that the basic tools of the geographer were eyes to observe and well-shod feet to cover the area being studied, observing, sketching, photographing and engaging the locals in friendly conversation (Figure 4). Despite his very tolerant approach in relative terms to the approaches chosen by his followers, he was wary of overly theoretical or indirect studies based on unchallenged statistics or models which were not verified at a local level. He gradually accumulated about 60 valuable fieldwork notebooks full of notes and drawings (a list of which can be viewed along with a bibliography on the site www.orlando-ribeiro.info. They are preserved in the archive of the National Library of Portugal). One of the notebooks relating to Portuguese Guinea was recently published in 2011. The photographic archive of the Centre for Geographical Research in Lisbon

Figure 4: Orlando Ribeiro teaching on an excursion. Photograph courtesy of the author.

holds several thousand of his photographs collected from 1937, only a small number of which have been published.

RIBEIRO'S SCIENTIFIC INFLUENCE AND LEGACY

Orlando Ribeiro left an original and deep-seated mark, in Portugal and the Portuguese-speaking world and in other countries too: as a lecturer in Portuguese language and literature in Paris and then as a professor of geography in various universities, mainly in Lisbon; as a researcher in areas ranging from Brazil to India, but mainly in the Iberian Peninsula and tropical Africa; and as Director of the Research Centre founded in 1943, which has recently achieved a European 'excellent' ranking (120 researchers, 52 of whom hold doctorates), and as an active member of the International Geographical Union and member of the Calouste Gulbenkian Foundation Science and Education Advisory Committees.

In Portugal, all geographers are still directly or indirectly followers of Orlando Ribeiro. When the issue arose in 1973 of finally establishing geography teaching in the Arts Faculty of the University of Porto, he was consulted and proposed to Rosa Fernanda Moreira da Silva, who developed an active teaching and research centre in a very short time in a turbulent revolutionary era. She was joined in 1982 by João Carlos Garcia, one of the most recent but closest followers of Orlando Ribeiro by virtue of his broad historical vision of geographical facts. Although Orlando Ribeiro's influence may appear to have waned over 30 years, when he retired and entered the 'hellish' period which always normally follows a long influential era of virtual hegemony, the announcement of his centenary in 2011 swiftly triggered a renewal of interest in Portugal. This has already facilitated work to disseminate the results of studies which began under his supervision and which continued after his death.

It is important, however, to note that Ribeiro's influence is not restricted purely to Portuguese geographic circles. For decades, he worked with the Portuguese Geological Service where he was involved in the surveying for several 1:50,000 maps. He also took part in the preparation of pamphlets on the climate in Portuguese regions published by the Meteorological Service. Over the course of 40 years of teaching in Lisbon, his lectures were attended by students from other disciplines in the Arts Faculty. Geography teaching was also offered in the Lisbon and Porto schools of Fine Art, initially by Ribeiro and then by his followers. By 1960, two of his direct followers, Francisco Tenreiro and Raquel Soeiro de Brito, were teaching at the newly created Higher Institute of Overseas Territories Social Sciences and Politics (*Instituto Superior de Ciencias Sociais e Políticas*). For several years, the Centre for Geographical Studies in Lisbon also maintained an active and close relationship with successive generations of alumni via monthly symposia where current research was presented and discussed. When Ribeiro's first book, *Portugal, o Mediterrâneo e o Atlântico,* was reprinted in 1962, it was considered a fundamental work on Portuguese culture, an essential and classic text. Strangely, but unfortunately, Orlando Ribeiro had little influence over the organization of geography teaching in Portugal at secondary level, although many extracts from his works were nevertheless included in textbooks. On several occasions, especially when he attempted to promote university teaching reforms, he broadened his audience by publishing a series of articles in newspapers. His clear and attractive style and his excellent grasp of teaching enabled him to reach a broad audience.

Ribeiro's influence abroad was significant and widespread, notably in Brazil, Spain, France and Italy. He was considered by his colleagues there to be a fellow countryman and a loyal friend to those countries. There were many mutual exchanges with France, far too numerous to mention in full here: they led not just to scientific collaborations but also to marriages. Among the French geographers working most actively in Portugal were Pierre Birot, author of major geomorphological research studies and also of an important synthesis of the Portuguese regions (1950); François Guichard (1946–2002), author of an important study on the sphere of influence of the city of Porto (1983); and the author of this essay, who worked mainly in Portugal on regional and historic geomorphology, climatology and geography. An excellent study of the rural history of the great plains of southern Portugal (1966) by Albert Silbert also had a profound influence on Ribeiro and was extended via the agrarian geography of Olivier Balabanian (1980). Two further French geographers based in Portugal are Denise de Brum Ferreira, author of an innovative thesis on the eastern Atlantic subtropical climate (1989) and Nicole Devy-Vareta, author in 1993 of a thesis on the little-studied topic of tree planting in the Serra da Cabreira.

There are also important links with Germany, the home of Hermann Lautensach, where two other followers, Ilídio do Amaral and Maria João Alcoforado, went to study following in the footsteps of Mariano Feio, and from where Bodo Freund came initially to carry out research and then to teach in Portugal. Other followers of Ribeiro went to pursue their studies in Sweden, England and the United States. There were particularly close links in the last named country with the University of Pennsylvania, courtesy of a former student in Paris, Gerald Moser, a specialist in African literature in Portuguese, and with Berkeley University, where Carl Ortwin Sauer (*Geographers* Vol. 2), whose concept of geography imbued with history was close to that of Ribeiro, and who was succeeded by the Brazilian geographer Hilgard O' Reilly Sternberg. In Canada, Ribeiro was a visiting professor in Quebec for a term in 1962, at the invitation of Professor Hamelin, and formed a firm friendship with the botanist Pierre Dansereau.

Orlando Ribeiro published little in English and is therefore still not very well known internationally outside Romance language countries. It is hoped that the present biography will partially remedy this and help make a selection of his works accessible to those who are not familiar with them via publication in what has become the contemporary international *lingua franca*.

Acknowledgements

The editors acknowledge with grateful thanks the role of Ian Thompson, Hélène Lamérant and Marianne Chalmers for effecting the translation of this essay from the French language original provided by Professor Daveau.

Bibliography and Sources

1. SELECTED BIBLIOGRAPHY OF WORKS BY ORLANDO RIBEIRO

A *Scientific Bibliography of Orlando Ribeiro*, containing 413 items dating from 1934 to 2008, has been compiled and annotated by Ilídio do Amaral and is available in full on the website www.orlando-ribeiro.info. Only Ribeiro's major works are given below, with the date of the first edition and the most recent edition.

1937 'A Arrábida. Esboço geográfico', *Revista da Faculdade de Letras* IV (1–2), 51–131 (2004, Third edition, Fundação Oriente, Lisbon: Câmara Municipal de Sesimbra).

1945	*Portugal, o Mediterrâneo e o Atlântico.* Coimbra: Coimbra Editora (1993, in a work illustrated with photos by Jorge Barros, Edições João Sá da Costa; 2011, Ninth edition, Lisbon: Letra Livre).
1949	*Le Portugal Central.* Lisbon: Congrès International de Géographie (1982, Second edition, Lisbon: Centro de Estudos Geográficos).
1949	*L'île de Madère.* Lisbon: Congrès International de Géographie (1985, Portuguese translation, *A Ilha da Madeira até Meados do Século XX*, Lisbon: Instituto de Cultura e Língua Portuguesa).
1954	*A Ilha do Fogo e as suas Erupções.* Lisbon: Junta de Investigações do Ultramar (1997, Third edition, Lisbon: Comissão Nacional para as Comemorações dos Descobrimentos Portugueses).
1955	*Portugal.* Volume V of *Geografía de España y Portugal.* Barcelona: Montaner y Simón.
1961	*Geografia e Civilização. Temas portugueses.* Lisbon: Centro de Estudos Geográficos, Lisbon (1992, Third edition, Lisbon: Livros Horizonte).
1962	*Aspectos e Problemas da Expansão Portuguesa.* Lisbon: Junta de Investigações do Ultramar.
1964	*Problemas da Universidade.* Lisbon: Livraria Sá da Costa.
1968	*Mediterrâneo. Ambiente e Tradição.* Lisbon: Fundação C. Gulbenkian (2011, Third edition also by this publisher).
1970	*Ensaios de Geografia Humana e Regional.* Lisbon: Livraria Sá da Costa.
1970	*Variações sobre Temas de Ciência.* Lisbon: Livraria Sá da Costa.
1973	(in collaboration with Suzanne Daveau) *La Zone Intertropicale Humide.* Paris: Armand Colin.
1975	*Destinos do Ultramar.* Lisbon: Livros Horizonte.
1976	*A Universidade em Crise.* Lisbon: Edições Cosmos.
1977	*Introduções Geográficas à História de Portugal. Estudo Crítico.* Lisbon: Imprensa Nacional, Casa da Moeda, Lisbon.
1981	*A Colonização de Angola e o seu Fracasso.* Lisbon: Imprensa Nacional, Casa da Moeda (2011, Second edition also by this publisher).
1985–6	(in collaboration with P. Birot and S. Daveau) *Les Bassins de Lousã et d'Arganil. I. Le bassin sédimentaire, II. L'évolution du relief.* Lisbon: Centro de Estudos Geográficos.
1986	*Iniciação em Geografia Humana.* Lisbon: Edições João Sá da Costa.
1987	*Introdução ao Estudo da Geografia Regional.* Lisbon: Edições João Sá da Costa (2001, Second edition also by this publisher).
1987–91	(in collaboration with H. Lautensach and S. Daveau) *Geografia de Portugal, I. A Posição Geográfica e o Território, II. O Ritmo Climático e a Paisagem, III. O Povo Português, IV. A Vida Económica e Social.* Lisbon: Edições João Sá da Costa (republished in part in various forms up to 2004).

1989–95 *Opúsculos Geográficos, I. Síntese e Método, II. Pensamento Geográfico, III., Aspectos da Natureza, IV., O Mundo Rural, V., Temas Urbanos, VI. Estudos Regionais.* Lisbon: Fundação C. Gulbenkian.

1994 *Finisterra.* Coimbra: Encontros de Fotografia.

1999 *Goa em 1956. Relatório ao Governo.* Lisbon: Comissão Nacional para as Comemorações das Descobertas Portuguesas.

2003 *Memórias de um Geógrafo.* Lisbon: Edições João Sá da Costa.

2. SELECTED STUDIES RELATING TO ORLANDO RIBEIRO

A list of 72 references, covering the period from 1964 to 2008, is available on the website www.orlando-ribeiro.info.

Alegria, M. F., Daveau, S. and Garcia, J. C. (2011), *Leite de Vasconcelos e Orlando Ribeiro. Encontros Epistolares (1931–1941).* Lisbon: Museu Nacional de Arqueologia, Imprensa Nacional, Casa da Moeda.

Amaral, I. do (1973), 'O Centro de Estudos Geográficos de Lisboa', *Finisterra* VIII, 310–15.

Amaral, I. do (1983), 'Estudos de Geografia das Regiões Tropicais (Contribuições da 'Escola Geográfica de Lisboa')', *Garcia de Orta* 8 (1–2), 1–44.

BNP (2011), *Orlando Ribeiro (1911–1997), Ponto de Partida, Lugar de Encontro* (Catalogue de l'Exposition). Biblioteca Nacional de Portugal, IGOT.

Bosque Maurel, J. (1998–9), 'El Professor Orlando Ribeiro (1911–1997) y su Obra Cientifica', *Boletin de la Real Sociedad Geográfica* 134–5, 325–31.

Cabo Alonso, A. (1998), 'Orlando Ribeiro, maestro de geógrafos', *Estudios Geograficos* 232, 375–88.

Daveau, S. (2007a), 'Planeamento e Difusão das Obras de Orlando Ribeiro', in *Geophilia. O Sentir e os Sentidos da Geografia.* Lisbon: Centro de Estudos Geográficos, 171–83.

Daveau, S. (2007b), 'La Géographie, ça peut survivre à la guerre. Correspondance entre Jean Gottmann et Orlando Ribeiro', *Finisterra* XLII, 5–21.

Ferro, G. (1998), 'Orlando Ribeiro (1911–1997)', *Bollettino della Società Geografica Italiana* XII, 1–5.

Garcia, J. C. (1998), 'Orlando Ribeiro (1911–1997), o Mundo à sua procura', *Revista da Faculdade de Letras da Universidade do Porto – Geografia* XIV, 107–16.

Medeiros, C. A. (1998), *A Geografia no Sentir de Orlando Ribeiro.* Lisbon: Departamento do Ensino Secundário, Ministerio da Educação.

Rebelo, F. (2008), *A Geografia Física de Portugal na vida e Obra de quarto Professores Universitários, Amorim Girão, Orlando Ribeiro, Fernandes Martins, Pereira de Oliveira.* Coimbra: Minerva.

3. OTHER WORKS CITED

Amaral, I. do (1964), *Santiago de Cabo Verde. A Terra e os Homens.* Lisbon: Junta de Investigações do Ultramar.

Amaral, I. do (1968), *Luanda. Estudo de Geografia Urbana*. Lisbon: Junta de Investigações do Ultramar.

Balabanian, O. (1980), *Les Exploitations et les Problèmes de l'Agriculture en Estrémadure Espagnole et dans le Haut Alentejo*. Braga: Barbosa & Xavier.

Birot, P. (1950), *Le Portugal*. Colin: Paris.

Capel, H. (1975), 'La Geografia Española tras la Guerra Civil', *GeoCritica* I(1), 1–25.

Devy Vareta, N. (1993), *A Floresta no Espaço e no Tempo em Portugal*. Faculdade de Letras da Universidade do Porto.

Feio, M. (1998), *As Causas do Fracasso da Colonização Agrícola de Angola*. Lisbon: Instituto de Investigação Científica Tropical.

Girão, A. de Amorim (1941), *Atlas de Portugal*. Coimbra.

Girão, A. de Amorim (1941–3), *Geografia de Portugal*. Porto: Portucalense Editora.

Guichard, F. (1992), *Porto, La Ville dans sa Région*. Paris: Fondation C. Gulbenkian.

Martonne, E. de (1925), *Traité de Géographie Physique* (Fourth edition). Paris: Colin.

Medeiros, C. A. (1976), *A Colonização das Terras Altas da Huíla (Angola)*. Lisbon: Memórias, Centro de Estudos Geográficos.

Mendes, C. (1985), *Maputo antes da Independência. Geografia de uma Cidade Colonial*. Lisbon: Memórias, Instituto de Investigação Científica Tropical.

Silbert, A. (1966), *Le Portugal Méditerranéen à la fin de l'Ancien Régime*. Paris: SEVPEN.

Spate, O. H. K. (1954), *India and Pakistan. A General and Regional Geography*. London and New York: Methuen.

Vidal de La Blache, P. (1922), *Principes de Géographie Humaine*. Paris: Colin.

Chronology

1911	Born 16 February in Lisbon
1916	Death of his mother; goes to live in the provinces with his maternal grandparents
1919	Returns to live in Lisbon in the Science Faculty district
1921–8	Student at *Passos Manuel* High School in Lisbon
1928–32	Reads History and Geography at the Arts Faculty of the University of Lisbon
1930–41	Becomes the friend and follower of the ethnologist Leite de Vasconcelos, whose works he publishes posthumously
1932–3	Attends lectures by the Swiss geologist Ernest Fleury, who teaches at the Higher Institute of Technology
1933–6	Teaches at the *Colégio Infante de Sagres*, in Lisbon

1934–7	Awarded a small bursary by the *Junta da Educação Nacional*, which allows him to travel throughout Portugal
1935	Takes part in the *Cruzeiro de férias* Atlantic cruise
1936	In March, presents doctoral thesis on Arrábida; marries Maria Chambers Ramos in November, they go on to have five children
1937–40	Portuguese language teaching assistant at the Sorbonne in Paris
1938	Takes part in the International Geographical Congress in Amsterdam
1939–40	Works as an assistant to Emmanuel de Martonne in Paris
1940	Returns to Portugal; first trip to Spain (*Congresso Luso-Espanhol* in Saragossa)
1941–3	Part-time lecturer at the Arts Faculty of the University of Coimbra
1943–81	Professor at the Arts Faculty of the University of Lisbon
1943–	Foundation of the *Centro de Estudos Geográficos*, of which Ribeiro is Director until 1970; host to Pierre Deffontaines in Lisbon; fieldwork with Hermann Lautensach and plans a *Geography of Portugal* with him; Pierre Birot settles in Portugal to work
1944	Hosts de Martonne in Portugal
1945	Publication of his best-known work, *Portugal, o Mediterrâneo e o Atlântico*
1946	Travels to Spanish Morocco and to Andalucia; takes part in the *Curso de Geografía General y del Pirineu* in Jaca; is appointed *Vogal da Junta de Investigação Colonial* and offered the Chair of *Estudos Olisiponenses*, created by Lisbon City Council
1947	Research trips to Portuguese Guinea and Madeira; travels to Paris; takes part in the Second International West African Conference in Bissau
1948	Research trip to Madeira; lectures on African Geography at the *Escola Superior Colonial*
1949	Leading role in the International Geographical Congress in Lisbon; leads excursions in Central Portugal and in Madeira; appointed Vice-President of the International Geographical Union; gives speech on the need for reform of the university system and scientific research in Portugal; in the *Colóquio* à la *Junta de Investigações do Ultramar* takes a stance on scientific research in overseas territories
1950	Undertakes trips to Europe and participates in the annual meeting of the British Association for the Advancement of Science in Birmingham
1951	Scientific meeting in Cairo for the 50th anniversary of the *Société de Géographie*; undertakes trips to Greece and to Italy; takes part in the Inter-University Pyrenean Excursion; travels to the Cape Verde Islands to conduct research and film the volcanic eruption on Fogo
1952	Research in west Africa and Third International West African Conference in Fernando Pó (followed by a research trip to the islands of Príncipe and São Tomé); undertakes extensive American tour before and after

	the International Geographical Congress in Washington; appointed First Vice-President of the IGU; first doctoral thesis by a follower, Mariano Feio
1952–3	Second research trip to the Cape Verde Islands
1953	Lectures at the Sorbonne (Paris), and in Strasbourg; first research trip to the Azores
1954	Travels to England; *A Ilha do Fogo e as suas Erupções* published but not distributed until 1957
1955	Publication in Barcelona of *Portugal*, Volume V of the *Geografía de España y Portugal*; second doctoral thesis by a follower, Raquel Soeiro de Brito, on the island of São Miguel in the Azores; takes part in an IGU Symposium in Kampala (Uganda)
1955–6	Research trip to Cairo, Karachi and Portuguese areas of India
1956	Travels to Bordeaux; takes part in the 18th International Geographical Congress and *Curso de Altos Estudos Geográficos* in Rio de Janeiro; first Honorary Doctorate at the University of Brazil in Rio de Janeiro
1957	Oversees the creation of a specialist Geography degree; receives Lopes do Rego Prize for his book on Fogo; appointed *Chevalier de la Légion d'Honneur*; travels to the Azores to study the Capelhinos eruption on Fayal
1958	Further trip to the Azores
1959	Lectures at the *Collège de France* (Paris) on the geography of Portuguese expansion; attempts to reform the Arts Faculty at the University of Lisbon, of which he has been appointed Dean
1960	Takes part in the 19th International Geographical Conference in Norden, including excursion to central Sweden; research trip to Mozambique and Angola and gives speeches on racial issues
1961	Second research trip to Angola and Mozambique with Fernandes Martins and Suzanne Daveau; third doctoral thesis by followers, Francisco Tenreiro, on the island of São Tomé; beginning of Ribeiro's involvement with the Scientific Advisory Committee of the Gulbenkian Foundation and its funding for the Centre for Geographical Studies
1962	Visiting lecturer Laval University, Quebec; trip to France; speaks out during unrest at Lisbon University; research trip to Angola with Pierre Gourou; Honorary Doctorate from Bordeaux University
1963	Further research trip to Mozambique with Pierre Gourou
1964	Lectures at the *Collège de France* (Paris); trip to the Alps; fourth doctoral thesis by a follower, Ilídio do Amaral, relating to the island of Santiago (Cape Verde Islands)
1965	Trip to Brazil; marries Suzanne Daveau
1966	Trip to Mexico; first issue of the biannual journal *Finisterra. Revista Portuguesa de Geografia*
1967	Organisation of *Primeiro Seminário Internacional de Geografia*, in Lisbon

1967	Visiting lecturer at the *Institut de Géographie*, Sorbonne (Paris); takes part in the Pan-African Congress on Prehistory in Dakar; research trip to Senegal and Mauritania
1968	First edition of *Mediterrâneo. Ambiente e Tradição*
1970	Trip to Spain and Morocco
1970–4	Preparatory work on the reform of university Social Science teaching with the Education Minister, Veiga Simão
1971	Awarded the Almágia Prize for his book *Mediterrâneo*; admitted to the *Accademia dei Lincei* (Rome)
1973	Publication of *La Zone Intertropicale Humide* (Armand Colin, Paris), in collaboration with Suzanne Daveau; further trip to the Canary Islands
1974	'Carnation' Revolution in Portugal
1976	Publication of *Destinos do Ultramar*, a collection of articles on Portuguese-speaking countries
1979	*I Colóquio Ibérico de Geografia* in Salamanca
1980	*II Colóquio Ibérico de Geografia* in Lisbon
1981	Celebration of his academic jubilee (retires aged 70); publication of *A Colonização de Angola e o seu Fracasso*
1983	Admitted to the Lisbon Academy of Sciences, in the Science section; draws up a will bequeathing his scientific archives to the Lisbon Centre for Geographical Studies
1984	First edition of a bibliography of his works by A. and I. do Amaral; Honorary Doctorate from the University of Coimbra
1984–8	*Livro de Homenagem a Orlando Ribeiro* (2 volumes, 99 contributions)
1985	Honorary Doctorate from the University of Madrid; trip to the Cape Verde Islands at the invitation of their President
1986	A brain haemorrhage brings his creative life to a close
1987	Appointed *Grande Oficial da Ordem Militar de Sant'Iago da Espada*
1987–91	Publication, in four volumes, of the *Geografia de Portugal*, a thematic collection of the works of Lautensach (1932) in Ribeiro's original translation (1955), updated by Suzanne Daveau
1989	Honorary Doctorate from the University of Paris-Sorbonne
1993	Publication of an illustrated edition of *Portugal, o Mediterrâneo e o Atlântico* with photographs by Jorge Barros
1994	*Finisterra*: catalogue of an exhibition of photos by Orlando Ribeiro, in Coimbra
1995	Special issue of the *Jornal de Letras*, dedicated to Orlando Ribeiro
1997	Dies 17 November

1998	A street is named after him by Pinhal Novo Town Council (in memory of a study from 1949)
1999	Posthumous publication of his *Report on Goa*; a street is named after him by Lisbon City Council
2003	Publication of *Memórias de um Geográfo*, with an introduction by João Carlos Garcia; inauguration of the *Biblioteca Municipal Orlando Ribeiro*, in the renovated Telheiras district of Lisbon
2004	Launch of the first version of the website www.orlando-ribeiro.info
2008	Theme issue (XLIII, 85) of *Finisterra* devoted to Orlando Ribeiro
2010	Signature upon the agreement under which the Lisbon Centre for Geographical Studies entrusts the management of a section of the Orlando Ribeiro's scientific legacy to the Portuguese National Library (*Arquivo de Cultura Portuguesa Contemporânea* Department); screening on television channel RTP 2 of the film *Orlando Ribeiro, Itinerâncias de um Geógrafo*, by António João Saraiva and Manuel Carvalho Gomes; launch of the second version of the *Orlando Ribeiro* website; screening on television channel RTP 1 of the film *Além de Nós. Mudanças na Paisagem*, by Anabela Saint-Maurice; inauguration of the exhibition *Orlando Ribeiro. Ponto de Partida. Lugar de Encontro* at the National Library of Portugal and award of the *Prémio Nacional de Geografia Orlando Ribeiro* by the *Associação Portuguesa de Geógrafos*

Aimé Vincent Perpillou

1902–1976

Hugh Clout

Education, Life and Work

Aimé Vincent Perpillou was born on 24 January 1902 at Glanges in the Haute-Vienne *département* of the Limousin region of the Massif Central. His father, André, was a soldier and his mother, Marie (née Bancaud) looked after the household. His secondary education was at the Lycée Gay-Lussac in Limoges and the Lycée Henri-IV in Poitiers, his father's profession having required the family to relocate to western France. At Poitiers he prepared to sit the national competitive examination (*concours*) for entry to the École Normale Supérieure in the rue d'Ulm in Paris, the elite training institution for teachers at both *lycée* and university levels which he entered in 1923. While in Paris he also followed classes at the Institut de Géographie (University of Paris, La Sorbonne) delivered by Lucien Gallois (*Geographers* Vol. 24), Emmanuel De Martonne (*Geographers* Vol. 12) and Albert Demangeon (*Geographers* Vol. 11), the direct disciples of Paul Vidal de La Blache (*Geographers* Vol. 12). Fellow students at the ENS included Jean-Paul Sartre and geographer Jean Dresch (*Geographers* this volume) who would later specialize in geomorphology and development studies, and become a professor at the University of Paris, heading the Institut de Géographie from 1960 to 1970. Perpillou's earliest research work was in climatology that earned him a *diplôme d'études supérieures* (the equivalent of a Masters degree by research). Among his fellow students in Paris was Suzanne Demangeon, the professor's daughter, who completed her research degree in 1927 (Demangeon 1928). She would become his wife and mother of his son and three daughters. Perpillou's qualities as a teacher had been demonstrated by his success in the demanding *agrégation* examination in 1927 that qualified him to teach in *lycées* or universities in France; he came top of the history and geography cohort comprising candidates from across the nation.

With a year of military service behind him, during which he rose to officer rank in the artillery, he taught for a short time at the Lycée Gay-Lussac in Limoges where he had been a pupil. He was then selected to teach maritime geography at two institutions in the port city of Brest, the École Navale and the École des Elèves ingénieurs mécaniciens de la Marine, remaining there until 1939. During the 1930s, he developed two lines of research: the first involved investigation of coastal and submarine features around western Brittany from which several articles were published; the second took him back to the Limousin region where he examined terrain features and changing land use for his *doctorat d'état* that was completed in 1940. This proved to be both a sad and a challenging year for the Perpillou household, since Albert Demangeon died in July and Aimé moved to a new teaching post at the prestigious Lycée Henri-IV in Paris. Located in the shadow of the Sorbonne, this *lycée* trained very intelligent pupils who sought entry to the ÉNS and other *grandes écoles*. During the war years his involvement with the Société de Géographie de Paris intensified and in 1947 he became *secrétaire général* of that organization, a position he would hold for a quarter of a century. His familiarity with the French geographical community equipped him to write a valuable review of French geographical publication during wartime that was published by the Royal Geographical Society (Perpillou 1946). At the end of the war, Perpillou moved to a lectureship at the University of Lille in northern France that introduced him to teaching in higher education. In this respect, he followed in the footsteps of Albert Demangeon who had taught in the Institut de Géographie in Lille between 1904 and 1911 (Clout 2003a, b).

Aimé Perpillou remained there for only three years before the professors at the Institut de Géographie in Paris invited him to occupy the chair of economic geography that had been Albert Demangeon's between the wars. Indeed, the Dean of Faculty André Cholley (*Geographers* this volume) allocated Perpillou the office that his father-in-law had used throughout that time. With a photograph of Demangeon prominently on display on the wall, the room would serve thereafter as a sort of shrine dedicated to his memory. Perpillou spent the rest of his career at the Institut de Géographie. He taught various aspects of economic geography including commodity production, transport and changing land use. Many of his lecture notes on these topics were duplicated by the Centre de Documentation Universitaire for sale to undergraduates in Paris, the provinces and more widely in the francophone world. No fewer that 42 titles of this kind of 'grey literature' are listed in his bibliography (Fénelon 1977). Perpillou was a clear and efficient lecturer whose style was that of a scholarly headmaster rather than that of a charismatic leader (Pierre George (1909–2006) would assume that latter role in the Institut de Géographie during the 1950s and 1960s) (On George, see *Geographers* Vol. 29). Perpillou is remembered as a kind man who offered wise advice to his advanced students and doctoral candidates, visiting them in the field when they carried out their research. Always concerned for their well-being, he was compassionate at times of illness or personal misfortune.

Beyond the Institut de Géographie, Perpillou taught at the École Normale Supérieure at Saint-Cloud from 1944 to 1955, and at the ÉNS (rue d'Ulm), his *alma mater*, between 1948 and 1952. He was an occasional visitor to the École Navale at Brest where he discussed current economic and political matters. He served on the panel of examiners for the history *agrégation* (1944–55) and examined the work of aspiring entrants to the military college of Saint-Cyr (1930–9), the École Nationale d'Administration (1953–8) and the Institut National Agronomique (1962–70). As secretary of the Société de Géographie, reinstalled in its headquarters at 184 boulevard Saint-Germain in 1944, he arranged the public lecture series (averaging ten each year) and edited its bulletin, re-titled *Acta Geographica*, for a quarter of a century. During that period it grew from being a simple listing of lectures, financial accounts, medals awarded and library acquisitions to a more varied publication that also published research articles. Perpillou is remembered as a welcoming editor who accepted papers from foreign scholars and was proud of the improvements in design, format and content that he managed to introduce. Nonetheless, the Société de Géographie remained true to its founding traditions in the world of explorers and travellers and few of its members were prepared for the 'new geography' that started to arrive in France from Scandinavia, the United States and the United Kingdom during the second half of the 1960s.

At the age of 70, Aimé Perpillou retired from his chair at the Sorbonne in 1972 and finally agreed to become president of the Société de Géographie whose 150th anniversary celebrations he had masterminded in November 1971. He served as a corresponding member of other geographical societies based in Antwerp, Budapest, Edinburgh, London, Madrid, New York and Vienna, with his perfect command of English facilitating his numerous international connections. Within the domain of French professional geography, Perpillou had joined the councils of both the Comité national français de géographie and the Comité des travaux historiques et scientifiques, remaining in those posts for more than a decade. His honours included Chevalier de la Légion d'Honneur and Officier des Palmes académiques de l'Ordre du Mérite agricole. To the shock of his friends and children, he died suddenly on 12 May 1976, his wife Suzanne having pre-deceased him. His son, Jean-Louis (born 1931), continued the family tradition of holding a Chair at the Sorbonne, but in classical Greek philology rather than in geography.

Scientific Ideas and Geographical Thought

The career, indeed the whole life, of Aimé Perpillou was profoundly influenced by his teachers at the Sorbonne and especially by Albert Demangeon who produced a 'model' regional monograph on *La Picardie et les régions voisines* in 1905 and numerous articles on aspects of human geography, including rural settlement and agriculture. In addition, Demangeon had started to research the denudation chronology of the Limousin region about which he published some preliminary results but did not carry the project to fruition (Demangeon 1910). This topic was duly passed on to his son-in-law whose doctoral work was supervised by the eminent physical geographer, De Martonne. Perpillou successfully defended his thesis in 1940 that appeared as *Le Limousin: étude de géographie physique régionale* (Perpillou 1940a), complementing other regional monographs that elucidated the physical geography of France. Perpillou published a number of complementary papers on the geomorphology of the region and on earth-surface processes in upland areas (Fénelon 1997). In addition, he wrote reports on coastal features that derived from his years in Brest and he contributed the annual *Chronique océanographique* to the *Annales de Géographie* between 1936 and 1945.

It was, however, his minor thesis on the changing rural landscape of the Limousin that would greatly influence his subsequent research career. Its central theme was the reclamation of moorland and the resultant increase in the cultivated surface during the course of the nineteenth and twentieth centuries, thereby transforming the countryside of this upland area with harsh climatic conditions and poor acid soils requiring application of lime and fertilizers. Such improvements were possible once railway lines had been installed across this section of the Massif Central. The main source that Perpillou employed was the land-tax (cadastral) record assembled for every land holding in the country and summarized at the level of each *commune*. The levies due were calculated from the nature, quantity and quality of land use, beginning in the early nineteenth century (the *ancien cadastre*) and revised prior to World War I and at subsequent stages in the twentieth century. Using such material, Perpillou was able to quantify the land-use pattern of the Limousin both before and after the transformation achieved by moorland clearance during the railway age. As its title shows, his secondary thesis emphasized the importance of mapping land-use conditions as an integral part of understanding countryside change.

Once again, the influence of Demangeon was pivotal in this choice of topic and source material since he had already written about the poverty of rural life in the Limousin and had devoted his own secondary thesis to the utility of archival sources for geographical study (Demangeon 1905, 1911). Perpillou (1935) had also published an early paper of the land-tax registers and maps at a time when historian Marc Bloch was also evoking their utility for scholarly research. Once settled at the Sorbonne, Perpillou launched his major research project into the changing land-use balance of every local *commune* in France, with the aim of creating three composite maps showing conditions for the early 1800s, the start of the twentieth century and the middle years of that century. The task was enormous, since the data were only available in the three cadastral registers drawn up at different dates for each of the 40,000 or so *communes* that constituted France. There was no convenient, centralized source for work at this scale, although *département* summaries were available (Clout 1983). Normally, the cadastral registers had been written out in duplicate, one copy being held in the *mairie* of the relevant *commune*, the second being placed in the tax offices (or sometimes in the local archives) of the chief town of the appropriate *départements*. With repositories in over 80 *départements* to be visited, simply obtaining and transcribing the raw data was an enormous operation. Perpillou encouraged his advanced students, doctoral candidates and young colleagues to assist him in this painstaking task, and in some cases, also recruited volunteers in provincial universities. Together with his many helpers he collated the results for individual *départements*, or groups of *départements*, from which he published many case studies in a variety of journals and conference proceedings (Fénelon 1977, 10). As well as conveying the statistical story of land-use change, he was determined to produce regional and national maps, devising a distinctive calculation that he used to depict land-use combinations graphically by means of vertical bands of varying colour and width (Sachet 1975; Solle 1981).

Perpillou's office at the Institut de Géographie of the Sorbonne was the headquarters for this operation that received funding from the national research organization (Centre National de la Recherche Scientifique). The precious statistics, copied meticulously into exercise books (since this predated the age of the computer), were kept in a fireproof cupboard in that room where one of his assistants was normally busy at a drawing table preparing artwork for publication. Many of the regional atlases that were being produced at this time included three land-use maps derived from Perpillou's data sets. By contrast, publication of the national maps proved both difficult and costly to arrange. In due course, however, three full-colour maps at the scale of 1:1,400,000 appeared in 1971, 1978 and 1981, their publication being made possible by grants from the CNRS and the Société de Géographie de Paris and, in the final case, thanks to a donation from Perpillou's heirs. These maps and the related articles convey a compelling story of the transformation of rural France from a composite of polycultural areas in the early nineteenth century, with wide expanses of uncultivated land in many regions, to a mosaic of cultivated land interspersed with areas of pasture and woodland in the twentieth century. The construction of railways, main roads and canals had allowed a degree of agricultural and land-use specialization to take place that had not been possible previously. However, by the very intricacy of their vertical bands, the detail of these maps is rather difficult to assimilate. At the present time when paper-based maps are less appreciated than they were half a century ago, the project launched and achieved by Perpillou has sunk into oblivion, even, it would seem, in the University of Paris (Marie-Claire Robic, personal communication).

Perpillou's devotion to the ideas and writings of his father-in-law was expressed in several, perhaps more surprising, ways. When Demangeon died in July 1940, his third contribution to the great *Géographie Universelle* that had been conceived before World War I by Vidal de La Blache and Gallois, remained in draft form (Clout 2003a, b). On 1,116 handwritten pages, it presented the economic and human geography of France and, together with De Martonne's text on the physical geography of France, was to form the *pièce de résistance* of the whole series. Initially the volume on France had been allocated to Gallois, but he became overwhelmed with editing the series following the death of Vidal in 1917 and made no progress on his own volume. Demangeon's draft presented the country as it was before World War II and lacked bibliographic references and photographs. Perpillou worked through the whole manuscript, revising and to a minor extent up-dating his father-in-law's work, adding references and photographs, and substantially rewriting the chapters on Paris. Many years later he would even publish two previously unpublished manuscripts by Demangeon that might have been drafted for inclusion in the Géographie Universelle (Demangeon 1971a, b).

Perpillou would undertake a similar task on several other publications. First, he arranged for Demangeon's doctoral thesis on Picardy that had appeared in 1905 to be republished in 1974 with an introduction outlining how the region had changed over the intervening years including the devastation caused during both world wars, and an appendix containing several new maps. Second, in 1975 he republished Demangeon's political geography text, *Le déclin de l'Europe* (1920) that showed how the balance of political and economic power in the world shifted following World War I, paying particular attention to the growing strength of the United States and anticipating the economic modernization of Japan, China and other world regions. Once again, Perpillou added his own commentary on how the global power struggle had changed as a result of World War II and went on to advocate further strengthening of the young European Economic Community. European projects in the field of transport and economic development were of particular interest to him. In similar vein, he had assisted Paul Reclus (*Geographers* Vol. 16), nephew of Elisée Reclus (*Geographers* Vol. 3), in the revision and abbreviation of the great man's work *L'Homme et la Terre* for an abridged (but still 3-volume) edition that appeared in 1931 with an acknowledgement of Perpillou's editorial role.

Perhaps, Aimé Perpillou's ultimate act of devotion to his father-in-law was to write the comprehensive text on human geography that Demangeon had sought to do throughout his career but failed to achieve. It should be noted that his *Problèmes de géographie humaine*, assembled by De Martonne after Demangeon's death, was not an integrated text but simply a collection of previously published articles (Demangeon 1942). The Demangeon-Perpillou papers, donated by Perpillou's son and daughters to the Bibliothèque Mazarine in Paris in 1978, reveal that Demangeon had written copious notes on tiny scraps of paper covering numerous aspects of human geography (Clout 2003a, b). These may well have assisted him in preparing lectures or school textbooks that he often wrote in association with appropriate schoolteachers or university colleagues. In addition, Demangeon had drafted his ideal plan for a major text on human geography (Grandazzi 1941). During the 1950s and 1960s, Perpillou wrote the book that his father-in-law had failed to achieve, deviating only slightly from Demangeon's outline. Beginning with general principles on the relationship between mankind and the natural environment, it examined forms of adaptation to the environment in five major world regions, technical factors and stages in human emancipation (especially related to industrialization and trade), and finally human settlement in terms of distribution of population, towns, states and nations. The textbook was duly published in Longman's Advanced Geographies series in 1966, but this was the very moment when the traditional content and teaching of human geography was being challenged and would soon undergo transformation in the light of new concerns for quantification, theorizing and model building. Perpillou's *Human Geography* failed to have the impact that it would have achieved had it appeared a decade earlier. It seems that a French version was not prepared for publication. With the book completed, Perpillou wrote a very appreciative essay on the work and career of his father-in-law that appeared in a volume presenting the activities of 15 French geographers (Perpillou 1975). Ironically, it was published not long before his own death when his recorded residence was the apartment on boulevard Henri-IV, looking towards the Seine and Notre-Dame, that had been the Demangeon family home between the wars.

Influence of Work and Spread of Ideas

Aimé Perpillou was in every respect a 'classical' French geographer trained in Paris by the direct disciples of Vidal de la Blache and remaining faithful to the ideas and practices of Albert Demangeon, his father-in-law. He saw his role as that of an educator and, like Demangeon, wrote useful texts for students, teachers and pupils at various stages of their education. As a professor he was clear and comprehensive, displaying his belief in the unity of geography (physical, regional and cartographic), although he was able to focus on economic geography during his quarter century of teaching at the Sorbonne (1948–72). His personal interests in transport geography were developed through the work of some of his doctoral students who prepared theses on aspects of railway geography, air transport, maritime geography and ports. Perpillou's fascination for tracing changes in rural land use, as recorded in land-tax documents, reached fruition in the three national maps spanning a century and a half as well as in an array of articles.

It was, however, Pierre George (1909–96), professor of geography at the Sorbonne since 1948 (and a colleague of Perpillou's in Lille between 1946 and 1948), who embraced new approaches to economic, social and political systems at a variety of scales and who managed to attract a very distinctive 'school' of doctoral candidates (Bruyelle 1998). Ironically, it was he, rather than Perpillou, who steered his students towards analysing the cadastral record in exhaustive detail to elucidate the importance of ownership of land in the French countryside by urban residents. This was part of a bigger project to determine changing processes of urban influence on rural areas during the nineteenth and twentieth centuries. It was George who wrote large numbers of small books and textbooks that fashioned a 'modern' (but still pre-quantitative revolution) brand of human geography that many undergraduates and members of the general public found very attractive.

Without doubt, Aimé-Vincent Perpillou served his students and his discipline faithfully throughout his career. He bore the substantial responsibility of being secretary of the Société de Géographie for over 30 years and by a cruel stroke of fate enjoyed only a few months as its president. He is remembered by his immediate students and by teachers who used his school textbooks but, as the years pass, that cohort of geographers is disappearing and the collective memory is fading among members of the French academy.

Bibliography and Sources

1. REFERENCES ON AIMÉ PERPILLOU, HIS FAMILY AND WORK

Bruyelle, P. (1998), 'L'Institut de géographie de Lille: cent ans d'évolution', *Hommes et Terres du Nord*, 63–70.

Clout, H. (1977), 'Some sources for the historical geography of rural France in the 18th and 19th centuries', in H. Clout (ed.), *Themes in the Historical Geography of France*. London: Academic Press, 541–64.

Clout, H. (1983), *The Land of France 1815–1914*. London: Allen and Unwin.

Clout, H. (2003a), 'In the shadow of Paul Vidal de la Blache: letters to Albert Demangeon and the social dynamics of French geography in the early twentieth century', *Journal of Historical Geography* 29, 336–55.

Clout, H. (2003b), 'Albert Demangeon, 1872–1940: pioneer of *La Géographie Humaine*', *Scottish Geographical Journal* 119, 1–24.

Demangeon, A. (1905a), *La Picardie et les régions voisines*. Paris: Armand Colin.

Demangeon, A. (1905b), *Les sources de la géographie aux archives nationales*. Paris: Armand Colin.

Demangeon, A. (1910), 'Le relief du Limousin', *Annales de Géographie* 19, 120–49.

Demangeon, A. (1911), 'La Montagne dans le Limousin: étude de géographie humaine', *Annales de Géographie* 20, 316–37.

Demangeon, A. (1942), *Les problèmes de géographie humaine*. Paris: Armand Colin.

Demangeon, A. (1971a), 'Les provinces françaises et le problème d'une réorganisation régionale', *Acta Geographica* 3(7), 151–208.

Demangeon, A. (1971b), 'La formation de l'Ėtat français', *Acta Geographica* 3(8), 217–38.

Demangeon, S. (1928), 'L'approvisionnement de Paris en légumes et en fruits', *Annales de Géographie* 37, 97–121.

Fénelon, P. (1977), 'Hommage au Professeur Aimé Perpillou', *Acta Geographica* 3(31), 6–13 [with full bibliography of the works of Perpillou].

Garenc, P. (1977), 'Aimé-Vincent Perpillou, 1902–1976', *Acta Geographica* 3(31), 3–4.

Grandazzi, M. (1941), 'Les éléments de géographie humaine d'Albert Demangeon', *Annales de Géographie* 50, 66–7.

Sachet, M. (1975), 'Les problèmes de changement d'échelle dans la representation cartographique de l'utilisation du sol', *Acta Geographica* 3(23), 3–13.

Solle, H. (1981), 'L'utilisation du sol en France. Les cartes Aimé Perpillou', *Acta Geographica* 3(45), 1–25.

2. SELECTED BIBLIOGRAPHY OF WORKS BY AIMÉ PERPILLOU

1935 'Les documents cadastraux dans les études de géographie économique', *Bulletin de l'Association de Géographes Français* 13–14.

1940a *Le Limousin, étude de géographie physique régionale*. Chartres: Durand.

1940b *Cartographie du paysage rural limousin. Atlas et commentaires.* Chartres: Durand.

1946 'Geography and geographical studies in France during the war and the occupation', *Geographical Journal* 107, 50–7.

1966 *Human Geography* (translated by E. D. Laborde). London: Longman.

1975 'Albert Demangeon, 1872–1940', in P. George (ed.), *Les géographes français.* Paris: Comité des Travaux Historiques et Scientifiques, 81–106.

Chronology

1902	Born 24 January at Glanges, near Pierre-Buffières in the Limousin
1914–23	Secondary schooling at *lycées* in Limoges and Poitiers
1923	Entered the École Normale Supérieure (rue d'Ulm) in Paris
1928–30	Taught at the *Lycée* Gay-Lussac in Limoges
1931–9	Taught in Brest at the École Navale and the École des ingénieurs mécaniciens de la Marine
1936–66	Publication of a range of textbooks for children in primary and secondary schools
1940	Publication of doctoral theses: *Le Limousin: étude de géographie physique régionale* and *Cartographie du paysage rural limousin: atlas et commentaires* (Awarded the Prix Pierre-Félix Fournier of the Société de Géographie de Paris for these works in 1942)
1940–5	Taught at the Lycée Henri-IV in Paris
1945	Appointed lecturer in geography at the University of Lille
1947	Appointed *secrétaire-général* of the Société de Géographie de Paris
1948	Appointed professor of economic geography at the University of Paris
1948–67	Publication of numerous sets of lecture notes on aspects of economic geography
1965	Publication of *Human Geography*, translated by E. D. Laborde
1972	Retired from the chair of economic geography at the University of Paris (Paris-IV, La Sorbonne)
1975	Elected *président* of the Société de Géographie de Paris
1976	Dies in Paris, 12 May

Two Vidalians: Antoine Vacher 1873–1920 and René Musset 1881–1977

Hugh Clout

Among the disciples of Paul Vidal de la Blache (*Geographers* Vol. 12), the names of Jean Brunhes (*Geographers* Vol. 25), Emmanuel de Martonne (*Geographers* Vol. 12) and Albert Demangeon (*Geographers* Vol. 11) come readily to mind. However the cohort of young men who espoused Vidal's 'new geography' in Paris before World War I comprised other scholars who, for various reasons, did not acquire such long-lasting reputations. Antoine Vacher and René Musset were two such Vidalians. Born eight years apart, their life spans were profoundly different, with Vacher dying in his prime aged 46 and Musset enjoying a productive retirement to the age of 96. Both men wrote doctorates under Vidal's supervision but their regional monographs were very different in character. Musset delivered a thoroughly balanced study in remarkably clear prose in which aspects of both physical and human geography were explored with sustained rigour. By contrast, Vacher concentrated almost exclusively on physical geography, paying no attention to contemporary socio-economic circumstances. Musset was a perfectionist who laboured long and hard to refine his logic and his style, while Vacher missed deadlines and let errors slip through but dared to do the unconventional. His early death robbed French scholarship of a promising physical geographer. Both men participated in interuniversity geographical excursions, and both taught at the University of Rennes, acquiring varying degrees of expertise in the geography of western France. After Rennes, Vacher taught at Lille but his career was interrupted by World War I. Musset left his mark in the literature and in the lives of his students at Rennes and Caen but he remained little known beyond France. Letters received by Albert Demangeon between 1904 and 1917 afford glimpses into the lives of both of these Vidalians.

Antoine Vacher

1873–1920

Education, Life and Work

Antoine Louis Marie Vacher was born on 18 September 1873 in Montluçon in central France to a family of tailors. His paternal grandfather had been an artisan, and his maternal grandparents worked the land. Antoine excelled in his early schooling and at the age of 7 obtained a scholarship enabling him to study at the Lycée de Lyon for the subsequent decade. He passed the *baccalauréat* (school-leaving) examination in humanities in July 1890. At the age of 18 he won a second scholarship that took him to the prestigious Lycée Henri IV in Paris, prior to entering the École Normale Supérieure (ENS) in the rue d'Ulm that prepared the nation's future academic elite. He studied there from November 1895 to November 1899. At the ENS, Antoine was taught by Paul Vidal de la Blache and from October 1900 continued his geographical studies at the Sorbonne, where Vidal had moved as professor in 1899. In 1896 he obtained his first degree in history and geography, and seven years later would obtain another degree in science. For 18 months after January 1899, he taught at the Ecole municipale Lavoisier in Paris. In 1900, having reached the age of 27, he passed the fiercely competitive *agrégation* examination, qualifying him to teach at a high level in a *lycée* anywhere in France. He continued to study at the Sorbonne until July 1901 when he was appointed to a tutorship (*maître-surveillant*) at the École Normale Supérieure. He held that position during the next four years when he prepared his doctorate, using libraries and archives in Paris during the academic year and working in the field during vacations.

His topic was the regional geography of the Berry, a province located to the north of his home town. However, he had to take on various jobs during the holidays to earn money to give his parents whose tailoring firm had fallen into debt (Condette 2001, 71). Within a few years of passing his *agrégation* he was publishing short articles on the economic geography of Argentina, on rainfall in Tunisia and on irrigation in Egypt in the pages of the prestigious *Annales de Géographie* founded by Vidal and Marcel Dubois (*Geographers* Vol. 30) in 1891 (Vacher 1902a, b, 1905a). These texts were critical digests of published work but Vacher was also producing original material. In 1904 he published a pioneering article on the urban geography of Montluçon and in the following year reported on the hydrology of the valley of the Cher, and on the changing course of a section of the river Vienne, working on the latter project with his friend Joseph Blayac (Vacher 1904, 1905b; Vacher and Blayac 1905). While much of his fieldwork focused on measuring river flow and valley forms, Vacher also read widely in oceanography and reported on that discipline to readers of the *Annales de Géographie* (Vacher 1905c).

Letters sent by Vacher to Albert Demangeon demonstrate a firm but occasionally strained friendship between the two men (Clout 2003a, b). In September 1904, he wrote to say how much he missed his 'companion in the cradle' now that Demangeon had moved from the École Normale Supérieure to the University of Lille (Letter dated 12 September 1904). Early in the following year, Vacher reported that he had 'visited *père Vidal* on one of his bad days. I talked to him about my thesis and we exchanged some sharp words; he ran down physical geography and I argued back. By the end of March, I shall have finished the purely physical part; by the end of June I shall have tacked the "anthropogeography". I shall finish this year' (Letter dated 8 February 1905). Unfortunately, these targets would not be met, but this did not prevent Vacher being appointed in October 1905 to a lectureship at the University of Rennes in place of Emmanuel de Martonne who had moved to the University of Lyon. In a humorous letter sent to Demangeon just before Christmas, Vacher restrained himself from using the university's headed notepaper lest his friend complain that he was 'wasting the resources of the Republic' (Letter dated 14 December 2005). Beneath the handwritten heading, 'Rennes, Faculté des Sciences', he reported that he had set up the equipment left by De Martonne in a room of its own.

The friendship between Vacher and Demangeon became notably strained over their collaboration, together with Joseph Blayac and others, on a *Dictionnaire-manuel illustré de géographie* that Armand Colin publishing house planned to bring out in 1907. In November 1906, Vacher acknowledged his tardiness, noting 'Vacher is still the same; he makes promises that he does not keep' (Letter dated 18 November 1906). When the book duly appeared in the following year, Demangeon was shocked to read harsh reviews and apparently accused Vacher of sabotaging the project. In an unusually formal typed letter, Vacher acknowledged that he had failed to discuss some large towns in the regions allocated for him to cover, had not always used the most recent census figures, and allowed errors to appear in his text on Latin America (Letter dated 9 February 1907). However, he refused to acknowledge the term 'sabotage' since he had not planned to harm the book or to damage the reputation of his fellow authors. Unlike his usual endings, 'Amitiés de coeur' or 'A toi de coeur' (in heartfelt friendship), he concluded rather more formally on this occasion, 'Bien cordialement à toi, Antoine Vacher'.

The *Dictionnaire-manuel* had generated a fee for Vacher but had been a distraction from working on his thesis. Early in 1907, he reported that he was suffering from migraine almost every day (Letter dated 1 March 1907). By November, two and a half years after his anticipated deadline, he completed his doctoral work but this still needed the approval of Vidal (Letter dated 1 November 1907). If, however, Vidal were too busy to read it then Lucien Gallois, his right-hand man at the Sorbonne, would do so (on Gallois, see *Geographers* Vol. 24). Vacher was proud to announce that Armand Colin, the prestigious publisher of Vidal, would produce the main thesis, subject to the master's approval. Vacher delivered his manuscript to the Sorbonne on 10 November 1907 but three months later Vidal had still not sent any comments. Instead, he dispatched a note to his disciple saying, 'Do not rush me. I like to enter into the thought of those whose work I read' (Letter from Vacher dated 5 February 1908). In exasperation, Vacher remarked, 'I shall not protest any more, I have to accept my fate'.

By mid-February 1908, Vidal had read the manuscript and simply requested that some passages be abbreviated and fuller captions be provided for the diagrams (Letter dated 16 February 1908). Vacher was now reassured. With his doctoral work complete, he felt able to marry a mathematics teacher, ten years his junior, who taught at the girls' *lycée* in Rennes (Letter dated 15 September 1908). As a man from humble beginnings, he confided to Demangeon: 'She has no fortune. I married her because she appeals to me' (Letter dated 15 September 1908). They would have only one child. Vacher's friend Jacques Levainville (1869–1932) reported, 'They made a fine couple', and commented that it was 'amusing to see this revolutionary anarchist thrown into the world of the bourgeoisie and clerics in Brittany' (Letter from Levainville to Demangeon dated 6 September 1908). Vacher then turned to academic matters, announcing that his doctorate would be examined at the Sorbonne in the third week of December (Letter from Vacher dated 15 September 1908; Clout 2009). All went well, with 'the boss being more friendly toward me than usual' and Lucien Gallois behaving 'just as you would expect the good man to do. Thanks to him, my thesis was graded *très honorable*' (Letter dated 19 December 1908).

With his doctorate finished, Vacher was promoted to the post of *professeur-adjoint* at Rennes. He threw himself into teaching and also accepted an invitation to organize the seventh inter-university excursion that would take place in Brittany early in June 1911. However, he complained to Demangeon about this extra burden that involved booking restaurants and hotels, and hiring a boat for part of the trip, as well as arranging its academic content (Wolff 2001, 335). The main objective was to study coastal geomorphology, beginning at the Baie du Mont Saint-Michel, then moving along the northern coast to Brest, and travelling by boat to the Ile d'Ouessant (Vacher 1912, 80). Three dozen students and academics attended from six universities, with Lucien Gallois (Sorbonne) and Pierre Camena d'Almeida (Bordeaux) being the most senior participants (on the latter, see *Geographers* Vol. 7). Physical features along the westernmost rias and the off-shore islands were inspected from the hired boat, being complemented by a visit to the Arsenal and a discussion of the commercial activities of the port of Brest by fellow Vidalian Camille Vallaux (*Geographers* Vol. 2) who taught at the town's naval academy. Seven years later, Vacher would publish a long article on the origin of these rias and Levainville, then employed by the Port of Brest, would write on commercial prospects (Vacher 1919). Vacher had a second excursion to organize during the summer of 1911 as part of the geographical 'pilgrimage' from southern Ireland to Italy that was being orchestrated by William Morris Davis (*Geographers* Vol. 5). De Martonne announced that he would not be available during the third week of August and volunteered the services of Vacher to undertake the daunting task of guiding Davis and his entourage in the field (Reported in a letter from Vacher dated 28 March 1911).

With this commitment over, Vacher prepared to move to the University of Lille to fill the teaching post that had been occupied by Demangeon prior to his appointment at the Sorbonne in November 1911. In the following January, Vacher was still looking for an appropriate house in the city, which he duly found in rue Kulhman, and was ordering additional books and maps for his students (Condette 2001, 68). In the summer of that year, he was rewarded for his contribution to Davis's 'pilgrimage' with an invitation to join 40 other European geographers on the great 'Transcontinental Excursion' to commemorate the sixtieth anniversary of the founding of the American Geographical Society (Clout 2004). This would run from 1 August to 5 October 1912, with Vacher being sponsored by the Ministère de l'Instruction Publique. Members of the French delegation contributed articles to the *Annales de Géographie* on the basis of their American expertise, with Vacher demonstrating his hydrological interests in an enthusiastic account of the Roosevelt Dam and irrigation systems in the Phoenix region

(Vacher 1913). He argued that physical conditions in Arizona, and notably the 'predominance of evaporation over precipitation, all serve to constrain the area assigned by nature to human activity, but none is capable of destroying human effort, especially when this is courageous and methodical. The example of Phoenix is the proof; it is also proof of the good work that the Reclamation Service has undertaken in the arid West' (Vacher 1913, 208).

On his return to France, Vacher devoted his energies to teaching at the University and at the École de Commerce in Lille (Condette 2001, 71). He also served on various committees of the powerful Société de Géographie de Lille, which brought together industrialists and merchants as well as academics and members of the general public, and he helped to edit its *Bulletin*. His health was not good, however, and in July 1914 he informed Demangeon that chloroform had been used in the surgery he had just undergone (Letter dated 27 July 1914). With convalescence and the academic vacation in mind, he thought that 'in a few days' time I shall be able to get away from Lille. I shall go to Brittany and I shall call in when I am in Paris' (Letter dated 27 July 1914). When the Germans crossed the Belgian frontier in August at the start of World War I, Vacher and his wife were safe in western France; however he planned to return to give his courses at the start of the new academic year in October. On 29 September, he wrote to the University authorities: 'My state of health is quite satisfactory for me to assume a more active life that I have here [in Brittany] . . . If the trains are not delayed too greatly, and if the Compagnie du Nord (Northern Railway Company) will accept us without prior reservation, we hope to reach Lille . . . Please excuse any delay in our arrival that will be due to circumstances independent of our will' (quoted in Condette 1999, 237). By the middle of October he was in the capital, as Lucien Gallois relayed to Demangeon on 17 October: 'Yesterday I had the pleasure of seeing Vacher who, together with his wife, arrived from Lille after a remarkable odyssey that he will certainly tell you about in better times. He is not getting on too badly' (Letter from Gallois, dated 17 October 1914).

Like a number of other Vidalians, Vacher spent part of World War I as a member of the Commission de Géographie producing geographical intelligence reports for the Service Géographique de l'Armée. His contributions related to Switzerland, the central Alps, the area around Venice, and Bavaria, and to the French *départements* of Aisne, Pas-de-Calais and Somme. Théodore Lefebvre, Vacher's pupil from Lille, wrote to Demangeon from various military hospitals and joked that thanks to the work of the Vidalians 'Germany must be almost completely conquered by the Service Géographique de l'Armée', while we [in the Allied armies] 'have not managed to cross the river Deûle' running through Lille (Letter from Lefebvre to Demangeon, dated 25 May 1917). With the Armistice signed in November 1918, Vacher provided cartographic assistance at the subsequent peace negotiations, and this enabled the Geography department in Lille to enrich its collection of maps and documents. The university had escaped material damage but the city had endured bombardment and four years of German occupation. On 24 April 1919, Vacher was awarded a personal professorship and he returned to his academic commitments, but he was granted leave from 16 November of that year because of ill health. On 1 January 1920, he was promoted to a full professorial chair, but the 'long and cruel illness' continued to take its toll and he died in Paris on 16 September at the age of 46 (Demangeon 1921, 70).

In a remarkably personal obituary, Demangeon reported:

> Up to the last days of his life, as long as he retained his physical strength and despite his suffering, we continued to find the close comradeship and friendship that had been forged through many years of working together; [he was] joyful and funny, full of curiosity and good spirit; honest, straightforward and good;

> concerned for others and glad to do the right thing. We saw death enter his soul and the illness gradually invade his whole body; he knew that he was fatally ill, but on his lips and in his facial expression we could still read the same confidence; seeing him so courageous, we were given fresh hope that life would conquer death and that he might return to work; but death claimed him in his prime and has not allowed him to complete his life's work. (Demangeon 1921, 70)

Scientific Ideas and Geographical Thought

Unusually for one of Vidal's pupils, Antoine Vacher was primarily a physical geographer whose expertise spanned hydrology, physiography and oceanography. He also wrote about several aspects of human geography, including the economic development of Argentina and the development of the port of Saint-Nazaire (Vacher 1902a, 1908c). The study of Montluçon was a pioneering example of urban geography; however Vacher chose to emphasize the town's physical setting in the valley of the Cher as well as the completion of the Berry canal in 1840 that allowed coal to be brought in from Commentry and permitted Montluçon 'to inherit, in the middle of the nineteenth century, the old metallurgical industry of Berry' (Vacher 1904, 131). He outlined the rise of other local industries and the growth of population from 4,600 in 1826 to 35,000 in 1901, but paid virtually no attention to the morphology of the town. He stressed 'the close ties that link mankind to the soil', maintaining that 'this relationship is discrete in a land like France: the land has been modified by Man from early times; the relationship varies according to historical circumstances, but it never disappears completely' (Vacher 1904, 137).

Vacher's main work was his doctoral thesis, *Le Berry: contribution à l'étude géographique d'une région française* (Vacher 1908a). This was accompanied by his minor thesis for the doctorate, *Fleuves et rivières de France: études sur les documents par l'administration des ponts et chaussées*, and a number of contributory articles dealing with the Cher, with river meanders, and with historic maps as sources for exploring fluvial geomorphology (Vacher 1905b, 1907, 1908b, 1909). The major thesis began in a conventional way with a definition of the region that Vacher equated with the 'land of the Bituriges' dating from Celtic times (Vacher 1908a, 1). As a good – if not unquestioning – Vidalian, he emphasized the 'collaboration between nature and mankind' and demonstrated how 'people changed natural conditions . . . by draining, by clearing woodland and by fertilizing farmland' across the centuries (Vacher 1908a, 7). His first chapter was, indeed, devoted to woods, heathland and marshes, while his second focused on the emergence of Berry as a historic province possessing a clear political boundary as well as a series of rather nebulous fringing landscapes. This point was further developed through an exploration of historic maps, dating from the middle of the sixteenth century to the late eighteenth century, as sources for geographical study.

Thereafter, Vacher's study adopted a radically different stance by focusing exclusively on aspects of physical geography, including geology and physiography, drainage patterns and valley morphology, climate, water supplies and river regimes, as exemplified by the Cher, Indre and Creuse. In conclusion, he returned to the issue of local *pays* names that had been reviewed in the first chapter and was close to the heart of Gallois at the Sorbonne. A large number of maps and photographs taken in the field confirm Vacher's detailed knowledge of his study area. His concluding remarks on 'generations of mankind in permanent struggle with nature' were classically Vidalian, however his almost exclusive emphasis on physical themes was not (Vacher 1908a, 490). Unlike his contemporaries, he paid virtually no attention to economic activities or population change. His archival research focused on information about river flow and flooding obtained from local branches of the Service des Ponts-et-Chaussées (bridges and highways authority), rather than on changing socio-economic circumstances. His minor thesis provided further information on the availability of archival sources for examining river regimes throughout France (Vacher 1908b). A long review of *Le Berry* by Emmanuel De Margerie, the Parisian physiographer and geologist, highlighted the originality of Vacher's approach and concluded that the rigour of his analysis could have earned the author the title of 'docteur-ès-sciences' rather than 'docteur-ès-lettres' (De Margerie 1909, 406).

Influence and Spread of Ideas

Despite the quality and undoubted promise of his work, the influence of Antoine Vacher was slight. He taught students at universities in Brittany and northern France but his short career was disrupted by World War I and the German occupation of Lille. His commitment to the work of the Commission de Géographie in Paris during the war afforded him little time to cultivate disciples, of whom some died in combat. A notable survivor was Théodore Lefebvre who would produce a doctorate and acquire professorial status before suffering the awful fate of decapitation at the hands of the Germans during World War II. Dying in the prime of life at 46, Antoine Vacher had no time to produce textbooks or exert an influence on the development of French geography between the wars. The life of this Vidalian was cruelly brief and destabilized the teaching of geography in the University of Lille. This responsibility passed into the hands of local *lycée* teachers until Maximilen Sorre (1880–1962) was appointed professor in 1922 (*Geographers* Vol. 17). He would remain at Lille until 1931, serving as dean of the faculty of arts for the final two years.

Acknowledgements

Letters received by Albert Demangeon between 1904 and 1917 are conserved in the Collection Perpillou-Demangeon at the Bibliothèque Mazarine in Paris. I am grateful to Professor Jean-Louis Perpillou, his grandson, for granting me permission to consult them.

Bibliography and Sources

1. *OBITUARIES AND REFERENCES ON ANTOINE VACHER*

Clout, H. (2003a), 'In the shadow of Paul Vidal de la Blache: letters to Albert Demangeon and the social dynamics of French geography in the early twentieth century', *Journal of Historical Geography* 29, 336–55.

Clout, H. (2003b), 'Albert Demangeon 1872–1940: a life in geography', *Scottish Geographical Journal* 119, 1–24.

Clout, H. (2004), 'Lessons from experience: French geographers and the transcontinental excursion of 1912', *Progress in Human Geography* 28, 1–22.

Clout, H. (2009), *Patronage and the Production of Geographical Knowledge in France*. London: Historical Geography Research Series, 41.

Clout, H. (2010), 'Jacques Levainville: a Vidalian by adoption', in R. Abad (ed.), *Mélanges Jean-Pierre Poussou*. Paris: Presses de l'Université Paris-Sorbonne, 1767–78.

Condette, J.-F. (1999), *La Faculté des Lettres de Lille de 1887 à 1945: une faculté dans l'histoire*. Lille: Septentrion.

Condette, J.-F. (2001), 'La création et le développement du Laboratoire de géographie de l'Université de Lille, 1898–1905', in G. Baudelle, M.-V. Ozouf-Marignier and M.-C. Robic (eds), *Géographes en pratique*. Rennes: Presses Universitaires de Rennes, 55–80.

Condette, J.-F. (2001), 'Les enseignants d'histoire et de géographie à la Faculté des Lettres de Lille sous la Troisième République, 1887–1940', *Revue du Nord* 83, 65–100.

Demangeon, A. (1921), 'Nécrologie: A. Vacher', *Annales de Géographie* 30, 70–1.

De Margerie, E. (1909), 'La géographie physique du Berry, d'après Monsieur Antoine Vacher', *Annales de Géographie* 18, 390–406.

Wolff, D. (2001), 'A travers les correspondances: l'envers ou l'enfer de l'excursion', in G. Baudelle, M.-V. Ozouf-Marignier and M.-C. Robic (eds), *Géographes en pratique*. Rennes: Presses Universitaires de Rennes, 327–42.

2. SELECTED BIBLIOGRAPHY OF WORKS BY ANTOINE VACHER

1902a 'La géographie économique de l'Argentine', *Annales de Géographie* 11, 247–59.

1902b 'Les pluies de Tunisie', *Annales de Géographie* 11, 269–71.

1904 'Montluçon, essai de géographie urbaine', *Annales de Géographie* 13, 121–37.

1905a 'L'irrigation pérenne en Egypte et les projets de Sir William Willcocks', *Annales de Géographie* 14, 80–3.

1905b 'Le haut Cher, sa vallée, son régime: étude d'hydrographie et d'hydrologie', *Annales de Géographie* 14, 399–423.

1905c 'L'océanographie moderne', *Annales de Géographie* 14, 289–95.

1905 (with J. Blayac) 'La vallée de la Vienne et le coude d'Exideuil', *Annales de Géographie* 14, 111–17.

1907 'La carte du Berry, par Jean Jolivet, 1545', *Bulletin de la Géographie Historique et Descriptive* 22, 258–67.

1907 (with A. Demangeon, J. Blayac, I. Gaillaud and J. Sion) *Dictionnaire manuel-illustré de géographie*. Paris: Armand Colin.

1908a *Le Berry: contribution à l'étude géographique d'une région française*. Paris: Armand Colin.

1908b *Fleuves et rivières de France: études sur les documents réunis par l'administration des ponts et chaussées*. Paris: Imprimerie Nationale.

1908c 'L'état actuel du port de Saint-Nazaire', *Annales de Géographie* 17, 271–6.

1909 'Rivières à méandres encaissés et terrains à méandres', *Annales de Géographie* 14, 311–27.

1912 'La septième excursion géographique interuniversitaire, 1911', *Annales de Géographie* 21, 80–3.

1913 'La région de Phoenix (Arizona) et le barrage Roosevelt', *Annales de Géographie* 22, 197–208.

1919 'La rade de Brest et ses abords. Essai d'interprétation morphologique', *Annales de Géographie* 28, 177–82.

Chronology

1873	Born on 18 September at Montluçon, son of a tailor
1890–1	Scholarship at Lycée de Lyon; passed *baccalauréat* exam in July 1890
1891	Scholarship to Lycée Henri IV, Paris
1895–9	Studied at the École Normale Supérieure (rue d'Ulm)
1896	Obtained degree (*licence*) in history and geography
1899–1990	Taught at the Ecole municipal Lavoisier, in Paris
1900	Passed competitive examination (*agrégation*) to teach in *lycées*
1900–1	Continued studies at the Sorbonne from October 1900 to July 1901
1901–5	*Maître-surveillant* at the École Normale Supérieure
1903	Obtained degree (*licence*) in science
1905	Appointed to lecture (*professeur-adjoint*) at the University of Rennes
1908	Married a teacher of mathematics in Rennes (late August); doctorate awarded in December; publication of *Le Berry*, and of *Fleuves et rivières de France*
1909	Appointed *professeur-adjoint* at the University of Rennes
1911	Organized the seventh interuniversity excursion in Brittany, 3–8 June; participated in the 'pilgrimage' from Wales to Italy organized by W. M. Davis (1 August–5 October); in November, appointed to teach at the University of Lille
1911	Participated in the Transcontinental Excursion across the United States, organized by W. M. Davis (22 August–17 October)
1914	Underwent major surgery in Lille during July
1914–15	Served in Paris as member of the Commission de Géographie for the Service Géographique de l'Armée
1919	Promoted to personal professorship at Lille, 24 April, but took extended leave on medical grounds
1919	Promoted to full professor, 1 January; dies, in Paris, 16 September

René Musset

1881–1977

Education, Life and Work

René Musset was born on 30 September 1881 at Melun (Seine-et-Marne), 45 km south-east of Paris (Journaux 1977, 6). His father was a headmaster who originated from Burgundy and changed posts frequently. The family moved several times when René was a boy with his schooling being undertaken in the Vienne, the Nord and then in secondary schools at Langres and Epinal in eastern France. The final phases of secondary education were completed at the Lycée de Nancy in Lorraine and the Lycée Louis-le-Grand in Paris. In 1902, René was admitted from this prestigious base to the elite École Normale Supérieure on the rue d'Ulm in Paris that trained the nation's high-flying teachers and many future university professors. He also attended classes by Paul Vidal de la Blache (*Geographers* Vol. 12), Lucien Gallois (*Geographers* Vol. 24) and Emmanuel de Martonne at the Sorbonne (*Geographers* Vol. 12). Musset received his first degree (*licence-ès-lettres*) in 1903 and two years later he came second in the fiercely competitive national examination (*agrégation*) that qualified him to teach history and geography at a high level in French *lycées*. For a year he taught at the *lycée* in Brest before receiving a three-year bursary from the Fondation Thiers (a graduate residential institution in Paris) that enabled him to start research under the supervision of Vidal. In 1905, Musset participated in the first interuniversity geographical excursion that was led by De Martonne in Brittany and presented the physical geography of the Rennes Basin and of the northern and southern coasts of the peninsula to two dozen students and professors from ten universities (Musset and Letaconnoux 1905–6).

Between 1909 and 1919, Musset taught geography and history at the Prytanée (military academy) of La Flèche in western France that served as a convenient base for his doctoral research on the nearby Bas-Maine area, effectively the Mayenne *département*. Initially he had intended to compare that area with conditions in the Perche but that plan was rejected (Letter from Musset to Albert Demangeon, dated 17 May 1906). Nonetheless, several articles on the Perche would be published after World War I (Musset 1919, 1920a). In November 1911 Musset's name was mentioned for a teaching post at the University of Rennes following Antoine Vacher's move to Lille. Vacher wrote to his close friend Albert Demangeon in the following terms: 'I have exchanged letters with Gallois and De Martonne about someone who might succeed me at Rennes. Gallois supports Musset. De Martonne shares my view that this would not be a happy choice. I think that [Henri] Baulig would be much better placed here, even though he has not completed his degree, and failing him then [Pierre] Denis . . . Despite having many reservations, I must confess that [Camille] Vallaux [at the naval college in Brest] seems to me to be superior to Musset who is a fine chap and a hard worker but for the moment has produced only a minor thesis' (Letter from Vacher to Demangeon, 30 November 1911) (on Baulig and Demangeon, see *Geographers* Vol. 4 and Vol. 2, respectively). So Musset remained instructing military officers at La Flèche, beyond the university domain. In 1917, he received his doctorate from the Sorbonne, with his major thesis, *Le Bas-Maine*, being produced by Armand Colin, the favoured publishing house of the Vidalians (Musset 1917a). His minor thesis on horse rearing in France appeared from a less prestigious publisher (Musset 1917b). Western France and aspects of agricultural geography would remain important components in his geographical repertoire for the rest of his life.

In 1919, René Musset was appointed professor of geography at Rennes following the departure of Baulig for the University of Strasbourg, which served as a showcase for French higher education after half a century of German occupation in Alsace. Musset would remain at Rennes for 14 years. Like most of his professorial colleagues teaching geography in provincial universities in France at this time, he alone was responsible for delivering all branches of geography. His research and publication during the years in Brittany focused on many aspects of western France, including two papers on the physiography and place names of Perche that emerged from his earlier investigations (Musset 1919, 1920a). In May 1919, he led the eleventh interuniversity excursion for professors and students of geography across the enclosed *bocage* landscapes of both Perche and Maine (Musset 1920b). An effective summary of much of his early work was published in English in the *Geographical Review*, dealing with both geomorphology and human geography (Musset 1922). Discussion of soil and climate was included under the latter rubric and emphasized 'human adaptation to natural conditions' (Musset 1922, 94). With the assistance of his friend Jacques Levainville (1869–1932) and of Professor Bigot, Dean of the Faculté des Sciences at Caen, Musset led a second inter-university excursion in May 1923 (Musset 1923a; Clout 2010). This centred on Caen and Cherbourg and involved up to 60 participants from seven universities in France and from two in Belgium, together with academic visitors from Princeton and Barcelona.

Musset continued to publish on many aspects of the geography of western France, including hydrology and climate as well as physiography and many facets of economic geography (Musset 1923b, 1929, 1934). All this was done in addition to preparing a text on Brittany for the Armand Colin collection edited by Demangeon. When Armand Rainaud, professor of geography at the University of Caen, retired in 1933, Musset decided to move to Normandy. His experience and authority made him an ideal candidate for election as Dean of the Faculté des Lettres four years later. He would bear that administrative responsibility, in addition to teaching and writing, until his retirement. *La Bretagne* was published in 1937 and would go through three revised editions (Musset 1937). Musset's fierce defence of the University of Caen and of Republican values during the German occupation (1940–4), together with a trumped-up charge of spying for the Allies, led to his deportation between May 1942 and April 1945, first to Compiègne, then to Sachsenhausen-Orianenburg and finally to Buchenwald (pers. comm. Pierre Brunet). In due course, he would accept the Médaille de la Résistance and become an Officier de la Légion d'Honneur (Journaux 1977, 7).

After the desperate experience of deportation, Musset returned at the age of 64 to Caen in the spring of 1945. He found the city centre in ruins following the Normandy landings of the previous June. The university had been destroyed completely and all his books, lecture notes and research documents were lost. With dogged determination that earned admiration from his colleagues and students, he continued to teach, write and administer, bearing an additional burden in the form of planning the renaissance of the university on a completely new campus site to the north of the great *château* built by William, Duke of Normandy, known in England as 'William the Conqueror'. At mid-century Musset responded to an invitation from Daniel Faucher (*Geographers* this volume), professor of geography at Toulouse, to contribute to a well-illustrated two-volume book entitled *La France, géographie, tourisme*, to be published by Larousse. Not surprisingly, his contribution covered western France and allowed him to present themes he had researched in earlier years (Musset 1952). He also placed considerable stress on the destruction of Normandy that had occurred in 1944 and the mighty challenge of reconstruction that followed.

Looking back half a century, Armand Frémont recalled Professor Musset in the following words:

> René Musset was a tall man, a real classic, an old man with a white beard but impermeable to age, possessing great personal culture and authority, unanimously respected because of his attitude during the war, when he had been deported. Working alone, he represented Geography [in higher education] at Caen and throughout Normandy. He knew the province well, having observed its physical and human aspects in detail. As a geographer, he favoured contact and observation, 'a geographer with his feet in farmyard manure' as he put it; but he was also competent with numbers, notably for demography and climatology . . . His typewriter and his calculating machine, which survived him by many years, would have merited a place in a museum of the rise of modernity . . . His little book, *La Normandie*, was one of my most solid sources of reference. (Frémont 2009, 13–14)

At the age of 73, René Musset took formal retirement, as the law in France then required university professors to do. Always a modest man, he disliked the events organized to mark this event, and continued to deliver one lecture course each year for the next three years (Journaux 1977, 7). In 1954 he became a founder member of the editorial board of *Norois*, the journal produced by a cluster of Geography departments in western France. With remarkable energy, he continued to write on the geography of Normandy, sometimes collaborating with his son Lucien (1922–2004) who was professor of medieval history at the University of Caen and an expert on the Vikings (Musset 1954, 1955, 1961). René Musset's book *La Normandie*, again for Armand Colin, appeared six years after retirement (Musset 1960). In addition to this geographical work, René Musset published numerous articles and notes for local journals. He was deeply committed to the learned societies of Normandy, including the Académie des Sciences, Arts et Belles Lettres (of which he served as president), the Société des Antiquaires de Normandie, and the Association Normande. He died in Caen in 1977. At 96, he was the last and the longest-lived Vidalian.

Scientific Ideas and Geographical Thought

Without doubt, René Musset was a devout follower of Vidal, whose 'new geography' was preached at the École Normale Supérieure and then at the Sorbonne, laying stress on the inter-relationship of physical and human phenomena and highlighting the importance of integrated regional enquiry. These founding principles were exemplified throughout Musset's career, in his doctoral thesis, his regional books and his research articles. *Le Bas-Maine* still stands as a 'classic' regional monograph *à la française*, whose comprehensive approach must have pleased Vidal and other professors at the Sorbonne (Gallois 1918). Why Musset did not dedicate his work to his mentor remains an enigma. He moved rapidly to discuss the *bocage* landscape that he saw as the predominant geographical characteristic of the Bas-Maine at the time of writing (Musset 1917, 1). However, in the early nineteenth century there had been an abundance of heathland in the area, and an extensive woodland cover prior to that (Musset 1917, 3). Musset argued that similar changes in the cultural landscape had occurred in other parts of western France, thereby rendering the Bas-Maine an ideal starting point to comprehend conditions across a much larger section of the country. He stressed that the Bas-Maine (the *département* of Mayenne) was not a well-recognized historic region, such as Normandy, Anjou and Poitou, but was simply an administrative fragment of western France.

Three main themes structured Musset's monograph: physical geography, human geography and 'the agricultural revolution' with its impact on the present landscape. Five introductory chapters ranged through discussions of erosion surfaces (in particular the 'peneplain of western France') to landforms, hydrology, climate and 'the influence of the soil and climate working in concert to determine river flow and to give rise to natural vegetation'. Of course, that vegetation cover 'experienced the profound impact of human action. One cannot talk of "natural vegetation"; the flora bears the imprint of human labour . . . Every trace of original vegetation has disappeared. The woodland that once covered all or almost all of the Bas-Maine has been cleared; today's forests have nothing in common with the primitive woodland cover; managed by man, they are in effect cropped' (Musset 1917, 177, 195). The 'human geography' that occupies the core of Musset's thesis is, in fact, a detailed account of the clearance of forests and the reclamation of heaths across the centuries, followed by a review of historic industries (iron-making, textiles, quarrying) and agricultural activities and marketing in the nineteenth century.

The 'agricultural revolution' that is explored in Musset's final section focuses on the diffusion of the practice of liming to enhance soil quality. All this was facilitated after 1830 by improved roads and after 1856 by railways that enabled anthracite and limestone to be united for the purpose of lime burning. In short, 'liming transformed the agriculture of the Bas-Maine', enabling soil to be transformed, encouraging heaths to be cleared, better cereal yields to be obtained and pastures to be improved for raising cattle, horses and pigs that were then dispatched to other regions for fattening (Musset 1917, 331). After lengthy discussion of agricultural changes, Musset dismisses industries, rural settlements, towns and population with remarkable brevity, and does not offer an overall conclusion to his work. Almost 100 maps and numerous photographs illustrate the text but it is strange that there is no map to show population change since the first official census of 1801. Musset's approach to morphogenesis has been overtaken by new ideas but the richness of his historical geography has stood the test of time (Brunet 1983, 10). Musset was clearly a competent all-rounder with a penchant for agricultural geography past and present. That preference was reinforced in his minor thesis on horse rearing and in subsequent texts that reviewed the production of wheat and rice on a global scale in the light of ecological and economic variables (Musset 1916, 1917, 1923c, 1944). It was Musset who produced the maps on livestock rearing for the first edition of the *Atlas de France* between the wars (Musset 1934, 1977). In 1951 he was honoured by election to the Académie d'Agriculture de France.

Musset's later books on Brittany and Normandy differ profoundly in structure one from the other. *La Bretagne* is thematic and examines a suite of characteristics from geology to population for the peninsula as a whole, with 'local aspects' being reviewed in a single chapter and only the territory around Nantes at the mouth of the Loire being singled out for distinctive treatment (Musset 1937). *La Normandie* recognizes a fundamental distinction between Basse- and Haute-Normandie (i.e. between western and eastern sections of the province) and then identifies a sequence of smaller *pays* whose physical, historical and contemporary characteristics are presented in a standard systematic way (Musset 1960). Musset's contribution to *La France, géographie, tourisme* evokes what he presented in the *Geographical Review*, with some themes and expressions reappearing 30 years later (Musset 1922, 1952). Essays written in his two decades of retirement confirm that he was the master of his chosen field, western France.

Influence and Spread of Ideas

In his long life, René Musset produced work that informed students who were required to understand the geographical intricacies of western France. As Pierre Brunet, a later professor of geography at Caen, observed: 'No geographer in Normandy can forget him' (Brunet 1983, 13). He influenced generations of students in Rennes and especially in Caen but his sphere of influence was tightly circumscribed. Despite early involvement with interuniversity excursions, and unlike many of his contemporaries, he did not offer a field trip for delegates at the International Geographical Congress held in France in the summer of 1931. However, seven years later he presented a paper at the 15th Congress in Amsterdam that traced the functional relationship between town and countryside, quoting the work of Walter Christaller (Musset 1938). Musset did not write a major thematic or regional textbook, nor – with the exception of his article of 1922 – did he publish in other than his mother tongue. He delighted in attending meetings of learned societies (*sociétés savantes*) in Normandy, but he tended to avoid international conferences and travelled little beyond France. He took his responsibilities seriously and carried his long academic and administrative experience to the Comité Consultatif des Universités and the Commission Supérieure de l'Education on which he served (Journaux 1977, 7). In the middle 1950s, it was highly appropriate for him to review 'recent regional studies' written by French geographers, but he remained something of a 'solitary figure' positioned on the fringe of that community (Musset 1957; Brunet 1983, 13). That aspect of his personality may have shaped Antoine Vacher's assessment of the man back in 1911. Without doubt, he was a profoundly self-critical perfectionist who insisted on rigour and precision in everything he did and expected the same standards from others (Brunet 1983, 13). Nonetheless, his 'rather gruff manner concealed caring thoughts and great concern for those individuals he held in esteem' (Brunet 1983, 13).

Musset's limited impact on the discipline must be set in the context of the times in which he worked. Like most provincial professors of geography, for most of his life he taught alone, not having an assistant until appointments in the 1940s. Initially, this was Louis Poirier, better known as the distinguished novelist Julien Gracq, who was joined by Jean Dresch (*Geographers* this volume) to cover for Musset from 1942 to 1946 (Gracq 1992, 158). Thereafter, André Journaux joined Musset and would be his professorial successor. Musset held the position of Dean for seventeen years and endured three harsh years in deportation camps, only to find his life's work destroyed when he returned to Caen. Physically weakened but morally courageous, he was confronted with starting his professional life afresh and assuming a significant role in rebuilding the city's university. Only with these facts in mind, may the contribution of René Musset – the last Vidalian – be appreciated.

Bibliography and Sources

1. OBITUARIES AND REFERENCES ON RENE MUSSET

Baudelle, G., Ozouf-Marignier, M.-V. and Robic, M.-C. (eds) (2001), *Géographes en pratiques, 1870–1945*. Rennes: Presses Universitaires de Rennes.

Broc, N. (1993), 'Homo geographicus: Radioscopie des géographes français de l'entre-deux-guerres (1918–1939)', *Annales de Géographie* 102, 225–54.

Brunet, P. (1983), 'L'œuvre scientifique de René Musset, 1881–1977', in *105e Congrès National des Sociétés Savantes, Caen, 1980, Section de Géographie*. Paris: Comité des Travaux Historiques et Scientifiques, 9–15.

Chevalier, M. (1996), 'Les géographes français dans l'entre-deux-guerres', in P. Claval and A.-L. Sanguin (eds), *La Géographie française à l'époque classique, 1918–1968*. Paris: L'Harmattan, 15–26.

Clout, H. (2010), 'Jacques Levainville: a Vidalian by adoption', in R. Abad (ed.), *Mélanges Jean-Pierre Poussou*. Paris: Presses de l'Université Paris-Sorbonne, 1767–78.

Frémont, A. (2009), *Normandie sensible*. Paris: Editions Cercle d'Art.

Gracq, J. (1992), *Les Carnets du grand chemin*. Paris: Corti, 158–67.

Journaux, A. (1977), 'René Musset, 1881–1977', *Norois* 24, 5–10.

2. SELECTED PUBLICATIONS BY RENE MUSSET

1905–6 (with J. Letaconnoux) 'Excursion géographique interuniversitaire en Bretagne, 1905', *Annales de Bretagne* 21, 251–76.

1908 'Les limites de la culture de la vigne dans l'Ouest de la France', *Annales de Géographie* 17, 268–70.

1914 'L'élevage du cheval en Camargue', *Recueil des Travaux de l'Institut de Géographie Alpine* 4, 297–310.

1917a *Le Bas-Maine: étude géographique*. Paris: Armand Colin.

1917b *L'Elevage du cheval en France*. Paris: La Maison Rustique.

1919 'Le Perche, nom de pays', *Annales de Géographie* 28, 342–59.

1920a 'Le relief du Perche', *Annales de Géographie* 29, 99–126.

1920b 'La XIe excursion géographique interuniversitaire', *Annales de Géographie* 29, 461.

1922 'The geographical characteristics of Western France', *Geographical Review* 12, 84–99.

1923a 'La XVe excursion géographique interuniversitaire', *Annales de Géographie* 32, 542–4.

1923b 'A propos des grands ports de commerce bretons', *Annales de Bretagne* 35, 209–27.

1923c	*Le Blé dans le monde*. Paris: Berger-Levrault.
1929	'Le relief du Haut-Maine', *Annales de Géographie* 38, 305–29.
1934	'La formation du réseau hydrographique de la Bretagne occidentale', *Annales de Géographie* 43, 561–78.
1937	*La Bretagne*. Paris: Armand Colin.
1938	'Rapports fonctionels des agglomérations urbaines et la campagne', in *Union Géographique Internationale, Comptes Rendus du Congrès International de Géographie, Amsterdam, 1938*. Leiden: Brill, vol. 2, 94–100.
1944	*Le Riz dans le monde*. Le Mans: Monnoyer.
1952	'La France de l'Ouest', in D. Faucher (ed.), *La France, géographie, tourisme*. Paris: Larousse, vol. 2, 61–168.
1954	'Population urbaine et population rurale en Normandie', *Annales de Normandie* 4, 299–314.
1955	'La maison normande, du Pays de Caux au Bocage normand', *Annales de Normandie* 5, 271–87.
1957	'Les études récentes de régions françaises', in G. Chabot, R. Clozier and J. Beaujeu-Garnier (eds), *La Géographie française au milieu du XXe siècle*. Paris: Baillière, 187–96.
1960	*La Normandie*. Paris: Armand Colin.
1961	'Le nom de l'ancienne ceinture forestière du Pays de Caux', *Norois* 8, 321–7.
1977	'La première carte des races bovines en France', *Norois* 24, 193–8.

Chronology

1880	Born at Melun, 30 September
1902	Entered the École Normale Supérieure, rue d'Ulm, Paris; obtained *Licence* degree
1905	Obtained *agrégation* qualification; appointed teacher at Lycée de Brest
1906–9	Research scholar at the Fondation Thiers in Paris
1909–19	Teacher at the military academy, La Flèche
1917	Obtained doctorate from the Sorbonne; publication of *Le Bas Maine*, and of *L'Elevage du cheval en France*
1919	Appointed Professor of Geography, University of Rennes
1933	Appointed Professor of Geography, University of Caen
1937	Publication of *La Bretagne*
1937–54	Dean of the Faculté des Lettres, University of Caen

1942–5	Deportation: Compiègne (May 1942 to January 1943), Sachsenhausen-Orianenburg (January 1943 to January 1945), Buchenwald (January to April 1945)
1945	Returned to the ruined city of Caen
1951	Elected to the Académie d'Agriculture de France
1954	Member of founding editorial board of *Norois*; retired from the University of Caen
1960	Publication of *La Normandie*
1977	Dies, aged 96, in Caen

Jean Dresch

1905–1994

Hugh Clout

Born into an intellectual family, Jean Dresch was attracted to geography through mountaineering. His first visit to Morocco revealed something of the misery of the third world, an impression reinforced later when he went to tropical Africa. Ten years of teaching in Morocco and membership of the Communist Party strengthened his opposition to the colonial system. His doctoral work focused on the physical geography of mountains but also examined human activities in harsh upland environments. He was firmly committed to the unity of the discipline of geography and insisted that theories should be tested by personal enquiries and observations in the field. As a communist, Dresch was able to visit China, the USSR and East European countries that were otherwise closed to Western scholars during the Cold War. His teaching career at the University of Paris lasted for a quarter of a century, focusing initially on North Africa and then on geomorphology, with increasing emphasis on arid regions, desertification and development studies. Jean Dresch directed the Institut de Géographie for a decade and mentored a hundred doctoral students, many of whom went to teach geography in developing countries. He presented papers at International Geographical Congresses throughout his career and was President of the International Geographical Union, 1972–6. A long and active retirement was devoted to academic travel, scientific writing and campaigning on behalf of the oppressed.

Education, Life and Work

Born in Paris on 30 November 1905, Jean Emmanuel Dresch was the son of the academic Joseph-Émile Dresch (1871–1958) and of Juliette Dresch (née Reau). Joseph-Émile had studied at the elite École Normale Supérieure (ENS), taught German at several *lycées* in the provinces and in Paris (Lycée Lakanal) and then at the universities of Nancy (1906–9) and Bordeaux (1909–22), before becoming rector of the *académie* of Toulouse (1922–31) and then of Strasbourg between 1931 and 1938 (Ruiz 1997, 251; Mombert 2005, 109). In these posts, he had responsibility for all aspects of state education from primary schools to the university in these parts of France. His son, Jean, was surrounded by books and a deep familial respect for learning. He excelled at school and was admitted to the ENS in 1926, on whose admissions panel his father had served from 1910 to 1923. Jean was fascinated by philosophy and had intended to focus on that subject. He was also exposed to geography, however, through classes by Emmanuel de Martonne (*Geographers* Vol. 12) and Albert Demangeon (*Geographers* Vol. 11) from the Sorbonne (Dresch 1956a, 1975). Tall and thin but physically robust, Jean was an excellent sportsman who developed an interest in hill walking and mountaineering in the Pyrenees, which were easily accessible from his parental home in Toulouse. It was at the ENS that he first encountered communist ideas, expressed by his fellow student Jean Bruhat (1905–83), who would become a distinguished historian of the working class (Suret-Canale 1994, 132).

While maintaining roots in philosophy, the intellectual focus of Jean's scholarship duly shifted towards geography, and especially the study of physical landscapes as revealed by De Martonne. Recalling his youth some 60 years later, Dresch explained that his 'preference for physical geography was, for a large part, due to a lot of mountaineering and wanting to understand why I liked doing it' (Dresch 1984, 38). He had intended to prepare his *diplôme d'études supérieures* (equivalent to a Masters degree by research) on a Pyrenean topic, but, purely by chance, Demangeon obtained funding for a student to do some research in Morocco and asked Jean if he would be interested. Having never set foot outside France, apart from crossing the Pyrenean frontier into Spain, Jean rose to the challenge and spent four months of 1928–9 in Morocco (Dresch 1979, 16). That decision would shape the rest of his life by exposing him to the kind of grinding poverty from which his bourgeois existence had shielded him, and introducing him to the mountains of the High Atlas that would be the study area for his doctoral thesis. Following the advice of local geographer Jean Célérier (1867–1962), Dresch conducted research in the Moulay Idriss upland for his *diplôme* and soon afterwards published two articles about the area in the *Annales de Géographie*, the leading journal of academic geography in France at that time (Dresch 1930a, b; Raynal 1963; Baduel 1986, 10). He then prepared for the fiercely competitive examination (*concours d'agrégation d'histoire et géographie*) for high-level *lycée* teachers and university lecturers. Jean duly passed this rigorous examination in 1930 and set off on his first teaching assignment in the following year.

Through the good offices of Demangeon and Célérier, Jean had been offered a post at the Collège musulman within the great courtyard (*mechouar*) of the royal palace in Rabat, where he taught history and Latin as well as geography, and then at the state *lycée* in that city (Dresch 1979, 17; 1994, 20). His teaching duties were not very time-consuming and he had plenty of opportunity to undertake preliminary field enquiries, which gave rise to a monograph on the physical geography of the mountains of northern Morocco (Dresch 1933a). Working at great distance from his patron De Martonne, Dresch researched his major thesis on the physical geography of the High Atlas and – as was the requirement at the time – prepared a small complementary thesis which examined the ways of life (*genres de vie*) of local farmers and pastoralists. In this way his expertise across the broad field of geography would be demonstrated. He acknowledged that his 'relations with De Martonne were deep, trusting and influential [but] neither of us was very communicative; our spontaneity was dampened by a natural shyness and this no doubt attracted us to one another' (cited in Bataillon 1983, 122). Nonetheless, De Martonne urged Dresch not to abandon his research and arranged for Henri Baulig (*Geographers* Vol. 4) to dispatch a bibliography of work by American scientists in comparable environments in the western United States (Dresch 1986a, 22).

As well as teaching and research, Jean Dresch became involved with various groups on the political left in Morocco, and investigated the inequalities between French colonial occupiers and the peasants and manual workers of the protectorate. He started to make a number of enquiries into economic and social conditions with a view to producing a kind of atlas but, while 'lawyers, administrators and geologists were interested, my immediate colleagues, especially if they were geographers, shunned my efforts' (El Gharbaoui 1978; Dresch 1984, 47, 1994, 22; Chouiki 2008). Indeed, 'the space between [Jean Célérier in] the higher education college and a secondary school teacher was a difficult gap to cross', despite both Dresch and Célérier being products of the ENS (Naciri 1984, 312; Dresch 1986a, 22). The French resident-general did not view Dresch's social commitment in a favourable light, nor the articles that he published – albeit anonymously – during 1938 and 1939 in *L'Espoir*, the outlet of the nascent communist party in Morocco (Naciri 1979). At the start of World War II, Dresch enlisted as an officer in the Moroccan rifle brigade (*tirailleurs marocains*), was involved in the brief raid on Narvik, but returned to Morocco in August 1940, after the armistice was signed between France and Germany (Prenant 1995, 214; Tissier 1996a, 370). By this time, he had completed the theses for his state doctorate (*doctorat d'état*) and had submitted them for De Martonne to scrutinize. Once his mentor had given his approval, the revised typescript had to be submitted to a publisher in France since university regulations required the volumes to be printed before the public examination at the Sorbonne could take place. Postal communication between France and Morocco was so irregular that Dresch was not able to check the corrections from his proofs (Dresch 1941a, 709). However, his days in Morocco were numbered since he, his wife Andrée (née Berger) and their young children were expelled for political reasons in January 1941 by the French resident-general, Charles Noguès.

Jean passed his oral examination with acclaim but his political orientation gave rise to obstacles in occupied France as he set about finding a permanent teaching post. Indeed, he would write later that De Martonne, whose sympathies were not to the left of the political spectrum, had prevented one of three newly created university posts in colonial geography being offered to him (Dresch 1984a, 43; 1994, 21). During the years of German occupation and the immediate aftermath of war, Dresch was involved in various short-term teaching contracts. From 1941 to 1944, he taught at the Lycée Voltaire in Paris, and gave classes on the geography of North Africa at the Sorbonne during the 1941–2 academic session. From 1942 to 1946 he also taught at the University in Caen since the resident professor, René Musset (*Geographers* this volume) had been imprisoned, and deported by the Germans on a trumped-up charge of spying for the Allies. Dresch was joined in Caen by another temporary lecturer, the geographer Louis Poirier who was researching the geomorphology of western France but would become world famous as the author Julien Gracq (1910–2007) (Tissier 1996b). Travelling between Paris and Caen each week was indeed a dangerous activity, especially at the time of Allied bombing before and after the Normandy Landings of June 1944. Back in the capital, Dresch participated in the Resistance and was part of the committee of Rol-Tanguy at the liberation of Paris in August 1944.

Despite having had no field experience of tropical Africa, Dresch was invited in 1945 to join an investigation of labour migration in West Africa, whose report would be used as evidence in the preparation of legislation for the abolition of forced labour in French territories overseas. This visit gave Dresch first-hand experience of what was then called 'black Africa' (as opposed to *l'Afrique blanche* of colonized North Africa) and it initiated a stream of scholarly work on migration, urbanization, European investment and uneven development as part of his wider critique of colonialism. Further opportunities to make observations in Africa took him to the Belgian Congo and French Equatorial Africa in 1946, Tunisia in 1947, Morocco on several occasions between 1946 and 1948 and parts of French West Africa in 1948 (Anon. 1948, 406). By autumn 1946, he had been appointed to a personal chair of geography at Strasbourg (1946–7), where his parents were living, but he took a year's leave of absence in 1947–8 in order to teach courses on the geography of North Africa and of 'black Africa' at the École Nationale de la France d'Outre-Mer in Paris, which trained future colonial administrators. At the same time, this father of six children was lecturing on the Muslim world at the École des Langues Orientales Vivantes and on the Union française (France overseas) at the Institut d'Études Politiques. On 1 October 1948, he moved to the chair of geography and colonization in North Africa at the Sorbonne where previous post-holders had adopted an emphatically pro-colonial line. The irony of this title was profound since Dresch's critique of the colonial system in the Maghreb and elsewhere was unremitting. Eight years later, he moved into the chair of physical geography following the retirement of André Cholley in September 1956 (Prenant 1995, 215) (on Cholley, see *Geographers* this volume).

Thereafter, Jean Dresch became director of the Institut de Géographie at the Sorbonne (1960–70), while continuing to bear a heavy lecturing load and serving as mentor to an important cluster of doctoral students working not only on physical geography in Morocco and elsewhere in North Africa, but also on aspects of human geography in tropical Africa, in the Near East and even in east-central Europe and the USSR, which Dresch's political orientation allowed him to visit. During the 'social revolution' of May–June 1968, he even stayed in the Institut overnight on some occasions to ensure that the building was not broken into. At this time he also directed the Service de documentation et de cartographie géographique in Paris, which was funded by the Centre National de la Recherche Scientifique. Following the fragmentation of the University of Paris, he moved in 1970 to the new Université de Paris VII where he headed the 'Laboratoire Tiers-Monde-Afrique' with historian Catherine Coquery-Vidrovitch (Bataillon 2009, 75). This interdisciplinary research unit would be known later as 'Sociétés en développement dans l'espace et dans le temps' (SEDET). Dresch presided over the Comité National de Géographie in France (1966–72), was a vice-president of the International Geographical Union (IGU), and then its president from 1972 to 1976, with his term of office concluding after the 23rd International Geographical Congress (IGC) in Moscow. Subsequently, he played a major role in preparations for the 25th IGC that would meet in Paris in 1984 (Dresch 1984b). By this time, Dresch was widowed (Madame Andrée Dresch having died in 1978), had been retired from his university chair for nine years and was well into a long and very active period of retirement in which he continued to travel extensively, visit researchers in the field and write unceasingly (Suret-Canale 1994, 133). For his life-long contribution to geography, he was awarded the Grand Prix de la Société de Géographie de Paris in 1988 (Beaujeu-Garnier 1989). Jean Dresch died in the capital on 6 March 1994, at the age of 88.

Scientific Ideas and Geographical Thought

Jean Dresch not only believed in the unity of geography but practised that conviction in various ways throughout his career (Beaujeu-Garnier 1995). His earliest work on the Moulay Idriss upland embraced the physical geography of the humid environment of northern Morocco as well as its cultivated landscapes, settlements and important sanctuary of Idris I, founder of the Idrisside dynasty (Dresch 1930a, b). His doctoral thesis focused on the relief of the High Atlas but his minor thesis identified and mapped ways of life in the mountains. His method of enquiry was direct: 'Like a nomad, I went from village to village, from sheepfold to sheepfold, and I came to learn their history and the alliances between people in one valley and in another. I was received like a friend and [learned so much] over glasses of tea or a bite to eat, and during chance meetings by the side of the track' (Dresch 1994, 20). The beautifully written study of the Massif de Toukbal within the High Atlas, which he drafted with biologist and fellow mountaineer Jacques de Lépiney, was dedicated to 'the harsh beauty of this mountain chain and to the labours of its inhabitants' (Dresch and de Lépiney 1938; Dresch 1979, 39; Dusserre 2009). Under the guise of a guidebook for tourists, it provided a vivid account of cultivation, irrigation, pastoralism and traditional settlements as well as of the peaks, valleys and rugged slopes of mountain scenery. Wide-ranging reports on the manifestation of colonialism in Africa and accounts of economic development in China and the USSR, which were 'open' to Dresch and few other 'Western' geographers, were matched by technical discussions of desert geomorphology, pediments and other natural phenomena across the globe (Dresch 1955a, b).

His *Géographie des regions arides*, written well into retirement and drawing on the experience of a lifetime, emphasized physical parameters but also included a review of human activities in deserts. Dresch was also acutely aware of the geopolitical significance of harsh desert terrain (Dresch 1980a). When invited to address the annual conference of the Institute of British Geographers in 1965, he chose to speak not about physical matters but rather the 'Utilization and human geography of deserts' (Dresch 1966). As Ian Douglas recalled:

> Jean Dresch, highly respected for his contributions to arid-zone geomorphology, [also] wrote many commentaries on the way the political system in Morocco before independence led to social injustice, rural overcrowding of traditional pastoralists on unsuitable terrain, soil erosion and land degradation with consequent increases in poverty. The delicate interplay of cultures, internal and external economic forces, land tenure systems and ecological fragility is well presented in his [textbook on the] geography of arid lands. (Douglas 1986, 459)

Dresch remained a firm believer in the virtue of observations in the field which he captured in sketches, photographs and jottings in his well-thumbed notebooks. At the age of 83, he declined an invitation to attend the 26th IGC in Sydney as a distinguished past president of the IGU in order to accompany young colleagues to China and to fly in an ultra-light aircraft across the Takla Makan dunes and the Kunlun corniche in Sinkiang (Dollfus 1994, 1).

Dresch's 700-page thesis for his state doctorate, dedicated 'To Monsieur De Martonne, with profound gratitude', was structured in four unequal parts dealing with the origin of landforms, detailed descriptions of the mountains (the longest section), base levels and landscape formation under various climatic conditions past and recent, spanning from the glacial to semi-arid (Dresch 1941a). In an enthusiastic five-page presentation in the *Annales de Géographie*, De Martonne stressed the many practical hardships that had confronted 'this young researcher', including difficult physical access in the mountains, personal insecurity, absence of topographic maps, geological surveys that were merely at the reconnaissance stage and exposure to 'a morphological history more complex than one could imagine' (De Martonne 1943, 49). Nonetheless, with 200 maps and diagrams, 90 photographs and 6 fold-out maps;

> by its abundance and quality, the illustration is exceptional. This is a treasure house of geological sections, panoramas and block diagrams prepared in a personal style, translating relief features in a dramatic way through use of violent contrasts of light and shade. Then comes a veritable atlas of original maps [and] three maps at 1:200,000 scale that result from a complete revision of the existing reconnaissance maps, correcting the depiction of features and place names, and

> providing a revision or correction of all the published geological surveys. (De Martonne 1943, 49)

The thesis was 'exceptionally rich' and was judged to be 'a complete success, only achieved by the vigorous temperament and strong personality' of its author. As well as interpreting the evolution and appearance of the High Atlas, Dresch advanced ideas about the operation of geomorphological processes under varied climatic conditions that might be tested in other areas of the globe (Gendre 1943). The Alpine geographer Jules Blache (*Geographers* Vol. 1) also expressed praise for the thesis but felt that 'the study has more of a geological appearance than a geographical one. M. Dresch is especially interested in the problem of dating features, rather like a geologist dating rocks' (Blache 1943, 429). Whether the thesis was a work of geography or of geology, a vast amount of new information had been discovered and important explanatory hypotheses advanced by its author, not least those regarding the operation of processes in varying of conditions from high mountain climates to peripheral dry deserts. Dresch was indeed innovative in the emerging field of climatic geomorphology.

Having spent many months in the High Atlas, he was intrigued by the cultural landscape of the High Atlas as much as by the natural scene:

> The way the local people used the mountains was fantastic! The way that they knew how to master the problem of water through irrigation and the problem of soils by constructing terraces was admirable! Moreover, this struggle against nature was a permanent drama, since every year the terraces, the earth barrages and the irrigation channels had to be rebuilt . . . Successful occupation of the mountains for more than ten centuries was maintained by permanently renewed human effort. The local people had established a kind of equilibrium with the forces of nature. But that would not last! They were, in effect, outside the modern system of production and consumption; the result has been migration away from the mountains. (Dresch 1984a, 39)

The maps from Dresch's minor thesis, showing distribution of population, transhumance routes, collective pastures and irrigation systems in the High Atlas were, he believed, 'greatly appreciated by civilian administrators and officers responsible for indigenous affairs, since they pinned them on the walls of their offices and were interested in what they showed' (Dresch 1984a, 47).

By the age of 35, Jean Dresch had displayed his professional commitment to both physical and human geography, which would continue through the rest of his life when visiting doctoral students in the field, attending fieldtrips associated with international colloquia (such as the IGU and the International Union for Quaternary Research (INQUA)), or preparing technical notes for publication (e.g. Dresch and Birot 1966; Dresch 1970). His preferred methods embraced personal observations and descriptions of landforms through word or image, being convinced 'that one must always check theoretical ideas through personal contact with people and things; for a geographer, critique can only be guided by contact with live reality and landscapes' (cited in Bataillon 1983, 136). He maintained that in Morocco, during the 1930s, a physical geographer 'could still do useful work by walking kilometre after kilometre and recording elevations, armed only with a notebook, geological hammer, camera, tape measures, clinometers and a rucksack' (Dresch 1978, 51). At that time, topographic and geological maps were deficient, coverage of aerial photographs was incomplete and laboratories with specialist technicians were unknown in Morocco. Fernand Joly characterized Dresch's approach to geography in terms of detailed description, scrupulous analysis, fieldwork and a concern for tangible, concrete evidence (Joly 1979, 4).

Even if he urged quantification, modern analytical techniques and teamwork on his doctoral students, one suspects that he was more comfortable being an 'old style, non-quantitative, descriptive geomorphologist' with an insatiable scientific curiosity (Ron Cooke personal communication, 20 April 2010).

As the years passed and Dresch travelled throughout the world, his focus widened from high mountain terrain to surrounding pediments, plains and plateaux, and especially to the 'hyper arid, arid and semi-arid agro-climatic zones' that cover about one-third of the land surface of the planet (Mainguet 1984, 600). With field experience of the Sahara, the Middle East, Central Asia, China, Australia and the Americas, he was eminently equipped to advise on processes contributing to desertification and duly submitted evidence to the UNESCO investigation into arid lands, however he refused to attend the major conference at Nairobi in 1977, since 'such conferences are of little value [and] cannot resolve fundamental problems' (Dresch 1977c; 1984a, 48). Reviewing the *Géographie des regions arides,* Andrew Warren praised Dresch's 'wide field experience and his eclectic and up-to-date reading' that were put to good use in this 'very informative' text that was 'crammed with detail' (Warren 1984, 258). As he stressed, 'Professor Dresch relies on the power of his prose rather than on illustration [and] although the prose does have an attractive momentum, and the authority of considerable amounts of detail, it seldom tries to explain processes with any theory, to analyse others' findings critically, or to pose any of the questions that have yet to be answered about deserts, their plant and animal life and their people' (Warren 1984, 258). Despite his discussion of urbanization being biased towards Islamic and pre-Islamic cities, and treatment of the implications of oil extraction being extremely brief, 'the book is a pleasure to read, and when Professor Dresch does give his opinions, they are balanced and constructive. He is particularly sensible, and here stimulatingly critical, in the section on desertification' (Warren 1984, 259).

The second feature to transcend much of the work of Jean Dresch is his recognition of inequality and injustice, his condemnation of human exploitation and his rejection of colonialism in all its manifestations. He dated this response to his initial arrival in Morocco in late September 1928, which was his very first experience of the world beyond his bourgeois life in mainland France (Dresch 1979, 15). Disembarking from the boat that had brought him from home, he was impressed by the new buildings surrounding the docks at Casablanca that symbolized the benefits that colonialism had brought to the protectorate of Morocco. As soon as he entered the medina of the old city, however, he was confronted by another scene: 'On a doorstep covered with filth, I saw a sleeping child, barefoot and in tattered clothes, his eyes and open mouth covered with flies . . . These first images have never left my memory' (Dresch 1979, 15).

These very images influenced his decision to return to teach in Morocco in October 1931. His work brought him into classroom contact with the sons of the bourgeoisie of Morocco in the Collège musulman and with the children of European administrators and settlers in the *lycée.* He also went on 'long, solitary journeys amongst the peasants who, with respect to technology, seemed to display pre-capitalist forms of organization' (Dresch 1979, 17). First impressions were bewildering and he realized that he had 'to undertake the difficult task of shedding received ideas and to try to understand "the others", their often contradictory modes of production, their cultures and traditions, and their forms of social stratification that were being overturned' by the impact of colonial power, just a few years after the Rif War of 1925–6, which some saw as a 'symbol of resistance to imperialism' in the protectorate (Dresch 1979, 18). Finding 'the ideas of Lenin or Stalin' inadequate to understand the social situation in Morocco, he started to discuss the activities of the 'Jeunes Marocains', who were organizing the first political party in the protectorate (the Comité d'action marocaine), not only with his pupils but also with Moroccan and European adults on the political left (Dresch 1979, 17).

Despite the Communist Party being banned, Marxist ideas began to spread in the mid-1930s among socialists in Morocco, with clandestine groups meeting in Casablanca and Rabat in 1935–6, which received formal recognition after the Front Populaire was voted into power in France in 1936. Most Europeans in Morocco opposed the new regime in the homeland and strongly rejected requests for political reform in the protectorate. Personal rivalries and doctrinal differences prevented Moroccan communists from working together to press for independence, and relations with the French Communist Party proved difficult. Nonetheless, a new journal entitled *L'Espoir* (*Hope*) served as a mouthpiece for communism in Morocco, with over 30 issues appearing between 1 May 1938 and 12 August 1939. At the time of the German-Soviet pact on 23 August, the administration prevented the final issue being published. Dresch had contributed many short, anonymous articles to *L'Espoir* dealing with Moroccan themes that ranged from poverty, colonial exploitation of Moroccan workers at home and in France, to specific scandals and cases of imprisonment. Seven of his articles were reproduced 40 years later in *Un géographe au déclin des empires* (Dresch 1979, 23–38).

This collection also included his sensitive biography of Hubert Lyautey, former general-resident in Morocco, which he wrote during the war at the request of Charles-André Julien. Dresch and Robert Delavignette argued that Lyautey was 'conservative, if not reactionary [in temperament], but he might also be considered to be socialist because he nationalized phosphate working and let it be known that he was not hostile, in principle, to the independence of the Moroccan empire' (Dresch 1945a, 74). Nonetheless, he wished to keep resource exploitation within sustainable limits and was forced to leave Morocco by frustrated Europeans who favoured accelerated development (Lacoste 1978a, 7). From his personal experience, Dresch concluded: 'If one compares the situation in 1912 with that in 1925, one would declare a miracle to have happened. When Lyautey arrived, he likened Morocco to a sinking ship . . . When Lyautey left in 1925 . . . Morocco was unrecognizable and so dynamic that it provoked the admiration of the world' (Dresch 1945, 89; Naciri 1979).

In 1945, Dresch had an opportunity to observe colonialism in 'black Africa' when the General Inspector for Colonies in the government set up by General de Gaulle decided to ask two academics who had been involved with the Resistance to join an official enquiry into labour exploitation in the Côte-d'Ivoire (De Suremain 2004). One of these was Michel Leiris, an ethnographer with previous field experience in Africa, and the other was Dresch, the only geographer to possess a state doctorate on a 'colonial' topic and to have been involved with the Resistance. The fact that he had not yet been awarded a permanent university post made the opportunity to go to West Africa all the more attractive: 'I judged it useful to discover a new [part of the] world where I would participate in something useful; to investigate and to try to suppress an abusive system, as in North Africa. In short, I had the opportunity to move further along the path that I had already taken, albeit in a different context' (Dresch 1979, 144). The journey started from Paris on 26 January 1945, with the plane arriving at Dakar six days later after having made many stops along the way; then an amphibious aircraft took Dresch, Leiris and the French officials to the Côte-d'Ivoire. Dresch found that 'this slow flight at low altitude was the best way to accustom my eyes to the terrain' that differed greatly from the mountains of Morocco (Dresch 1979, 144).

Over two months spent in the colony were followed by a journey through the British Gold Coast and then back into the French territory of Guinée. Plantations, mines and forestry camps were visited and interviews were held with 'both black and white employers' and with migrant labourers about topics ranging from landholding, manpower needs, accommodation, costs of foodstuffs and health conditions in the various workplaces (Dresch 1979). Dresch explained: 'These enquiries, certainly rapid, were nonetheless a remarkable introduction to the depths of the geography of Africa for me' (Dresch 1979, 144). The empirical information he and Leiris had gathered was incorporated in the general report and was used to support the case for new legislation to improve the lot of migrant workers in French West Africa and in other French territories overseas. For many years, their account was mislaid in the archives of the former Inspection Générale des Colonies with no copy having been made, since 'in May 1945, carbon paper was as rare as typists in Dakar' (Dresch 1979, 146). The typescript was eventually retrieved, however, and a section of the report was reproduced in *Un géographe au déclin des empires* (Dresch 1979, 147–63). Dresch later declared that it was one of the most useful pieces of geographical work that he had ever undertaken (Dresch 1984a, 47).

His experience of colonial West Africa was broadened during later visits to many areas in French, British and Belgian territories. An intricate sequence of personal observations later supported by official statistics enhanced his discussion of labour exploitation, enabled him to evaluate the practice of rice cultivation and its potential for future development, and helped him to trace the diversity of urban life in tropical Africa, with some places demonstrating 'extraordinary vitality' and growth as a result of townward migration under colonial rule but others stagnating or declining (Dresch 1945b, 1949, 1950, 200). A basic distinction between planned 'white towns' (or districts) and overcrowded, poverty-stricken 'black towns' (or districts) was drawn and he then developed a simple typology with the addition of 'Soudanais towns', 'ports of Senegambia', 'towns of Guinée' and 'Congolese towns' (Dresch 1948a). More stimulating than such a loosely organized collection of field observations, was his sharply focused comparison of the old colonial centre of Brazzaville in French Equatorial Africa and the modern city of Léopoldville in the Belgian Congo, only 20 minutes away by boat across Stanley Pool (Dresch 1947a). Using these two urban places as the starting point for his article, Dresch proceeded to explore varying manifestations of European colonialism as expressed through administration, investment and economic development in two adjacent territories under different imperial powers.

His analysis of investments by colonial nations and economic corporations in their various overseas territories, plantations and factories proved to be a particularly innovative and instructive avenue of research (Coquery and Coquery-Vidrovitch 1980). An initial study revealed that British and Belgian investments in their respective tropical African territories substantially exceeded on a *per capita* basis those made by France in its colonies in West Africa and Equatorial Africa (Dresch 1946). To complement this international comparison, Dresch proceeded to focus on the French Union (as the imperial territories were then called) and added some historical depth by comparing interwar rates of investment with those being made immediately after the war (Dresch 1953a). Once again, 'L'Afrique noire' was very much the poor relation when compared with rates of investment by France both in North Africa and in Indo-China where levels of economic development were already more advanced (Dresch 1952).

Towards mid-century, Dresch refocused his attention on French territories in North Africa and published a substantial article on recent socio-economic changes in the *Annales de Géographie* (Dresch 1948b, 1953). Adopting a rather familiar form of organization, he compared the high productivity of colonial – often irrigated – agricultural estates with the growing misery of pastoralists and cultivators in the poorer environments of the Maghreb. Then he showed how rural poverty was fuelling waves of townward labour migration and demonstrated that shanty towns were sprawling around ancient urban cores and new 'colonial towns' alike, since urban housing was in short supply. Finally, he argued that modern industrialization, funded by French banks, financial corporations and development agencies, was being sited at locations that suited the investors rather than the inhabitants of the three territories. Much of this evidence, complemented by a thorough discussion of the fragile physical resource base, was refashioned in his 135-page discussion of North Africa in the textbook he wrote with Pierre Birot on *La Méditerranée et le Moyen-Orient*, whose final section bore the ominous subheading, 'The crisis of the colonial economy' (Birot and Dresch 1953, 513–14). Independence was indeed imminent in Morocco and Tunisia (1956) and the start of what would be the prolonged war of Algerian liberation (1954–62), that would see the departure of European residents from France's most favoured territory across the Mediterranean, was only months away. As early as 1956, Dresch published an article in the communist journal *La Pensée* in which he proclaimed that 'It will be the Algerians who will decide with the French what links may be wished to be maintained between France and Algeria; it will be the Algerians themselves who will specify their political status and the relationship between the two groups of population, of different civil status, who are established side by side on the land of Algeria' (Dresch 1956b, 124). This overtly political statement was remarkable for being made by an academic geographer and for being delivered before the French Communist Party declared itself in favour of Algerian independence (Tissier 1996a, 371). It represented just one element in Dresch's reaction to the decline of colonial empires across the globe and their replacement, he would argue with some dismay, by manifestations of postcolonial economic dependence on former colonial masters.

Influence and Spread of Ideas

Two interwoven strands characterize the career of Jean Dresch: his sustained belief in the unity of geography, and his commitment to the needs of the poor and exploited, which he expressed by his allegiance to communism throughout his adult life. His geographical ideas were conveyed to many audiences from *lycée* pupils and their teachers, and undergraduates, to doctoral researchers and professional members of the academic discipline. Unlike Pierre George, his colleague at the Sorbonne, Dresch did not write short books for pupils in the final classes of secondary school, university students and members of the general public. Instead, he expressed his ideas through articles in a surprisingly wide range of outlets, extending from scholarly geographical journals to others that either aimed at a wider readership (such as *Politique Étrangère*, *Revue Politique et Parlementaire* and *Présence Africaine*) or were associated with communist sympathies (e.g. *Démocratie Nouvelle*, *L'Espoir*, *La Pensée*). His involvement with writing school textbooks was brief and occurred in the early part of his career. Under the contributing editorship of André Cholley, he collaborated with René Clozier to produce an advanced text entitled *La France: métropole et colonies* (1936) for the J.-B. Baillière publishing house to serve a new syllabus in school geography. It was revised by the same authors and republished as *La France et l'Union Française* in 1948.

As a professor at the Sorbonne, Dresch taught the largest cohort of undergraduates taking geography courses in France and concentrated on the challenge of understanding North Africa and the intricacies of the world's landforms. He was a firm believer in comparing the situation in one region or environment with that in another, and then adopting a global view. As Claude Bataillon remarked, Dresch dealt with topics of immediate importance 'such as colonization in his teaching on North Africa. Combining his strong voice with sweeping hand gestures, he gave clear and vigorous lectures about societies and landscapes' (Bataillon 2006, 6). To Olivier Dollfus, these classes always adopted an analytical, critical tone and seemed to 'offer a pessimistic view of things' despite a constant quest for social justice and for a more reasonable balance between mankind and the natural world (Dollfus 1994, 2). Certainly Dresch, George and fellow geographer Jean Tricart were members of the Communist Party but they did not seek to indoctrinate their students. Rather, 'their convictions were expressed as a consensus of the left-wing, in which Cholley also participated. There was a consensual atmosphere [among these members of the geography staff] rather than an expression of hard militancy' (Bataillon 2006, 8). To complement his lectures at the Sorbonne – and in several other institutions of higher education in Paris, including the ENS – Dresch collaborated with Pierre Birot on the two-volume text, *La Méditerranée et le Moyen-Orient* in the Orbis series edited by Cholley for the Presses Universitaires de France (Birot and Dresch 1953, 1956). Dresch drew on his many years of living in Morocco in the chapters on North Africa, whereas the 300 pages that he wrote on 'The Arab Middle East' relied quite heavily on published information and on field observations made during short visits. While greeting the first volume with unqualified praise, reviewers offered a more nuanced reaction to the second volume dealing with the Middle East (Fisher 1954; Anon. 1956). Dresch's only other textbook was his *Géographie des régions arides* (1982).

Throughout the 1960s, Jean Dresch directed the Institut de Géographie at the Sorbonne and managed to retain the respect of a large and increasingly varied group of colleagues during most of that period through listening to all points of view and making every effort to act fairly. In the opinion of Olivier Dollfuss, his pupil and eventual colleague: 'His regime was marked by an openness of spirit, and by the exercise of power without favouritism or abuse' (Dollfus 1984, 80). As well as running the Institut, Dresch for several years chaired the national board of examiners for the *agrégation de géographie*, which determined the career progression of the most scholarly *lycée* teachers and young lecturers. He presided over the geography committee of the Centre National de la Recherche Scientifiique (CNRS), which allocated funds for research projects, laboratories, doctoral fellowships and publications. He was also chair of the Comité National de Géographie, which integrates the work of various systematic sub-sections of the discipline and acts as a sort of 'college' of French geographers in possession of doctoral qualifications. In addition, Dresch directed the École Supérieure de Cartographie Géographique and was founding head of the Service de Documentation et de Cartographie Géographique, which received funding from the CNRS (Dresch 1984b).

As well as these many administrative responsibilities, he found time to visit many of his doctoral students in their respective study areas in the Maghreb, the Middle East and parts of Latin America. A particularly important cluster of state doctorates on the physical and regional geography of Morocco was completed by René Raynal (1959), Fernand Joly (1960), Jean Le Coz (1964), Gerard Maurer (1968) and Gaston Beaudet (1969) under his patronage (Clout 2011). Focused on different environments from those of the High Atlas, these works greatly extended geographical knowledge of the country that had fascinated Dresch since the late 1920s. By contrast, André Prenant, Yves Lacoste and other protégés based in Algeria had to abandon or delay their original projects for state doctorates, as conditions worsened during the war of independence (Bataillon 2006, 13–14). Despite his best efforts to retain a sense of cohesion during the troubled months of 1968, the Institut de Géographie came to be divided among three of the dozen 'new' universities (Paris I, IV, VII) that replaced the old, monolithic and seriously overcrowded Sorbonne. To the dismay of some in Paris I, he decided to spend the final years of his career in the innovative institution of Paris VII where physical geography was represented strongly and there were many social scientists working on development studies. In the years prior to his retirement, he devoted much of his energy to graduate students. Indeed, he is said to have supervised about a hundred doctorates after 1960, including lengthy state doctorates as well as the shorter *doctorats de 3e cycle* (Dollfus 1994, 2). Many of the latter were submitted by postgraduates originating in North Africa, the Middle East, tropical Africa and other parts of what was then called the third world. By training so many scholars, he played a vital role in establishing or reinforcing geographical education in many developing areas and well as preparing experts who would work as economic or urban planners in their home countries (Lacoste 1978a, 8). As well as witnessing the fragmentation of staff, space and facilities in the Institut de Géographie, Dresch failed to have the building renovated and enlarged to house additional lecture halls, seminar rooms, laboratories, staff offices and an expanded library.

As chair of the Comité National de Géographie, Jean Dresch was the French delegate in the IGU. He had presented papers at almost every IGC since 1931, the first three of which derived from his research in the High Atlas of Morocco, with later communications relating to tropical Africa, arid lands and physical processes on desert fringes (Dresch 1933b, 1936, 1938, 1951, 1957, 1960a, b, 1972, 1977). At the 16th IGC held in Lisbon in 1949, he read a paper on arid and semi-arid pediments (Dresch 1951). Membership of the Communist Party meant that he was denied a visa to enter the United States to attend the 17th IGC at Washington, DC in 1952. His colleagues George and Tricart were excluded for the same reason; however his joint communication on French geographical research in tropical Africa and Madagascar was delivered by co-author, Charles Robequain (Dresch and Robequain 1957). Despite being excluded from the United States in 1952, he would soon prove to be 'an acceptable geographer in the East as in the West, and in the North as well as the South at the time of the Cold War' (Dollfus 1994, 2). In 1960, he was one of the chairmen of the geomorphology section of the 19th IGC at Stockholm, and four years later he contributed to discussions on denudation chronology at the 20th Congress in London. He was a particularly active founding member of the Commission on Present Day Geomorphological Processes at the 21st IGC at New Delhi in 1968 and later became chairman of that commission. He devoted 'a good deal of time to this work, leading to advances in international collaboration in physical geography' (pers. comm. M. J. Wise, 3 August 2010).

All this activity brought Jean Dresch to the attention of the Executive Committee of the IGU which decided at Budapest in August 1971 to nominate him for election by the general assembly in 1972 as President for the term 1972–6, which included the 23rd IGC held in Moscow in 1976. He continued to promote geomorphology at international scientific meetings and strove to increase the number of member nations belonging to the IGU: entry of the People's Republic of China would not be achieved until 1984 (Harris 1974, 30; 1996, 292). Chauncy Harris stressed that

> Jean Dresch was widely respected as a scientist. He was known to be active in communist affairs in France but the executive committee did not think that political affiliation should play a role in a scientific union. After all, the then president of the IGU Stanislaw Leszczycki was a member of the Polish communist party . . . Dresch was an excellent member of the executive committee. I clearly was not communist but worked with Dresch . . . very cordially and effectively. Politics was not on our agenda. (Harris 296)

Michael Wise insisted that 'his political stance (while one knew about it) was not obvious in his IGU work, it was his strong international scientific interest that counted' (pers. comm. M. J. Wise, 3 August 2010). Dresch represented the IGU on several other scientific bodies, including INQUA and the IGCP (International Geological Correlation Programme).

For Andrew Goudie and Michael Wise:

> Dresch was a leading figure in the IGU for two decades [and] became well known as an active member of the Commission for Arid Zones (1960–68). He developed

> strong views on how research should be organized at the international level of a Union with eighty members and was appointed Chairman of the Standing Programme Committee which reported during the 21st IGC in New Delhi in 1968. Dresch was ready to sweep away many of the products of history in favour of a reduced number of twelve effective Commissions. But, in a memorable and stormy meeting, his clear rationally-argued views failed to carry the support of the General Assembly which elected, in all, twenty-five Commissions. Dresch was, however, elected a Vice-President of the IGU and this led to the Presidency (1972–76) . . . As President, his tall lean figure and impressive forwardly projected head with its well marked features gave him a powerful presence. He dealt pragmatically with the many problems of holding a Congress in Moscow in 1976: his Presidential Address reflected his interests in the organization of research, in the world-wide communication of information, and in the development of geographical studies in the less developed world. (Goudie and Wise 1994, 359)

Despite being a foreign member of the Academy of Sciences of the USSR, Dresch failed to convince the Academy that foreign geographers attending the 23rd IGC in Moscow wished to have large scale topographic maps of the city region and of other areas they might visit on related excursions. As Chauncy Harris explained:

> Under regulations of the former military alliance of the Warsaw Pact no maps of a scale larger than 1:250,000 showing accurate longitude and latitude were to be available to the public. President Dresch, on behalf of the IGU, wrote to the Academy of Sciences noting that geographers would be hampered in their scientific work at the congress if they did not have such maps. [His attempt failed, since] the control of such maps rested with the security apparatus of the USSR. No action was taken, and geographers did not have good maps for the congress, but at least President Dresch made a strong case and put his scientific credentials on the line. (Harris 1996, 300–1)

When the question arose of publishing his presidential address in France, Jean Dresch chose to place it with *La Pensée* (Dresch 1977a, b). Under the interrogative title, 'Crise de la géographie?' he argued that recognition of crisis was a salutary sign of a wish to innovate, change and improve, since 'a discipline that did not experience a crisis would be feeble or dying' (Dresch 1977b, 20). He acknowledged the multiplicity of specialisms within the subject and agreed that no single scholar could be in command of them all. He insisted that a real challenge existed for geographical research to be undertaken collectively, with physical geographers, human geographers and colleagues from other disciplines joining together in teams to tackle the world's many pressing problems (Dresch 1977b, 25). This is exactly what he strove to achieve at Paris VII.

Professor Michael Wise served as a Vice-President of the IGU during Dresch's presidency. Many years later, he remembered the support that Jean had received from Andrée, his wife, and recalled a visit on IGU business:

> I went to Paris to meet him and he took me to his handsome apartment, lined with books . . . At first I found him rather austere. I was in awe of him. He had an established and very distinguished reputation, but I came to admire him greatly . . . He was President of the IGU at the difficult meeting in Moscow when there were diplomatic problems, especially over the attendance of the Israeli delegation . . . The troubles rumbled on during the Congress itself but I thought that Jean (with the support of Chauncy Harris then Secretary General) handled the

> problems superbly. He was calm and unflustered, always keeping the principle of internationalism to the fore. I greatly admired him – he was an example to me of how to deal with such problems. (pers. comm. M. Wise, 3 June 2010)

A further issue to be resolved in Moscow was the number of future specialist commissions, with 16 being proposed by the Executive Committee but a further 11 being suggested during the 14th General Assembly of the IGU. After 'a lively discussion', under Dresch's chairmanship, a total of 17 was agreed upon (Harris 1977, 10). Michael Wise duly succeeded Dresch as President of the IGU from 1976 to 1980, when Dresch retired from its Executive Committee at the age of 75. Chauncy Harris paid specific tribute to Jean Dresch as a distinguished scientist, effective administrator and active leader of the IGU. Firmly based in French geographical traditions, this multilingual scholar 'reached outward to the international geographic community' and earned 'profound respect and deep affection' (Harris 1996, 304).

Beyond the framework of academic geography, Dresch sought to establish scholarly links and to 'build bridges' across the world, always operating against the backdrop of his political allegiance. As early as 1952, he helped to found the Amitiés franco-chinoises to promote international friendship and to work towards the diplomatic recognition of China by France (Dresch 1984a, 48). He presided over that association for a dozen years and his academic visit to China in 1954 gave rise to a long article about the vast country that was closed to most Western scholars. However, his concluding remark, 'population is growing at an accelerated rate without yet creating any concerns: the child is king', would be overturned a couple of decades later, with the one-child policy being introduced in 1978 (Dresch 1955a, 201). Dresch was elected an overseas member of the Academies of Sciences of Poland and of the USSR and, together with Pierre George, travelled to Soviet Central Asia and other regions in 1955, which proved to be the first of several visits (Dresch 1955b, 1960c, 1968, 1986b). Colloquia between French geographers and their Russian or Polish counterparts owed much to his initiatives. He was also one of the leaders of the Association française d'amitié et de solidarité avec les peuples d'Afrique (AFASPA) (Suret-Canale 1994, 134).

Jean Dresch was one of the first French intellectuals to speak out against the harsh repression of Algerian demonstrators in Paris in October 1961, and four years later he denounced the seizure and disappearance of the activist Mehdi Ben Barkha, who had been one of his earliest pupils in Morocco and had become a friend (Tissier 1996a, 371). In response to a request from the Vietnamese Communist Party to investigate the American bombing of dikes that protected paddy fields along the Red River, Dresch requested his disciple Lacoste to undertake fieldwork with the International Commission of Enquiry into War Crimes in the first fortnight of August 1972 (Claval 2000, 244). Dresch then took advantage of his presence at the 17th IGC in Montreal to transmit Lacoste's findings to international journalists covering the meeting and to roundly condemn the attacks, arguing that the Americans had tried systematically to destroy the farming basis of the Vietnamese economy through their strategic bombing (Lacoste 1972, 1973; Clayton and Bowd forthcoming). In the eyes of Chauncy Harris, Dresch was right to stress 'the environmental damage resulting from the Vietnam war [since] environmental damage is a proper subject of concern for the IGU' (Harris 1996, 296).

Unlike many French academics who left the Communist Party after the Russian invasion of Hungary in 1956, Dresch stayed a card-carrying member to the end (Spire 1994). Remaining true to his geographical and ideological convictions, he was 'an exemplary and an exceptional *mandarin*' (Dollfus 1994, 2). Armand Frémont spoke for many when he recalled 'an immense admiration for the man and his work' (pers comm. A. Frémont, 24 July 2010). To Jean Suret-Canale, his communist colleague at Paris VII, he was 'a master, a comrade, and a friend' (Suret-Canale 1994, 134). Despite his many achievements and powerful positions,

> His high functions did not diminish his simplicity, or his fraternal availability for all those who called upon him. There was nothing 'mandarinal' in his behaviour, nor the slightest trace of pontificating pretension, which affects so many university people (and many others too) once they have acquired a certain reputation or achieved promotion in their career. He never presented himself as an 'oracle' or 'delivered lessons' to those seeking his advice. (Suret-Canale 1994, 134)

Conclusion

By virtue of the breadth of his intellectual grasp, his tenacious energy and his firm diplomacy, Jean Dresch enjoyed world renown among academic geographers and scholars working in cognate fields. In the 1960s and 1970s, when he was at the height of his powers,

> one could not be unaware of Jean Dresch. Not only was he involved in all the recruitment exercises [for geographers in French universities], but he was elsewhere too. His great silhouette was visible in China as in Latin America, in the USSR as in Africa. His voice rang out in the halls of international conferences, and in ordinary meeting rooms in the University of Paris, which he was not able to renovate. (Dollfus 1984, 73)

An issue of *Hérodote*, bearing the title 'Géographie/ anticolonialisme: Jean Dresch', appeared soon after his retirement, in which a dozen contributors evoked his profound influence on geographical work in north and 'black' Africa, his sustained critique of colonialism and his support for independence movements in Algeria and elsewhere (Lacoste 1978b). A volume of *Méditerrannée* brought together 16 articles in his honour, dealing with physical and human aspects of the geography of North Africa and major environmental and socio-economic challenges facing the Maghreb countries (Joly 1979). The festschrift published under the title *Désert et Montagne au Maghreb*, to commemorate Dresch's eightieth birthday, confirmed that his influence extended far beyond the already broad confines of his home discipline, since essays by geographers were outnumbered by contributions from agronomists, historians and social scientists (Baduel 1987).

Despite his frustration at the perpetuation of inequality and injustice across the face of the globe, even after the decline and formal disappearance of world empires and the creation of so many independent states, Jean Dresch continued to enjoy the practice of geography – in its fullest sense – through his final years. Fernand Joly declared: 'Throughout his career, Jean Dresch remained a physical geographer rather than a geomorphologist, and even simply a geographer rather than a physical geographer since he was interested in everything on the face of the Earth' (Joly 1979, 4). In letters to Jacqueline Beaujeu-Garnier (*Geographers* Vol. 28) Dresch explained:

> I am a specialist in physical geography and one might be surprised that I am also preoccupied with human problems. Experience shows that a physical approach is not a bad introduction to a global geography of underdeveloped countries . . . where terrain, climate and ecological conditions have a particular impact on land use and human activities . . . My reasons for loving my craft as a geographer?

> As a teacher certainly, in contact with young people and their many concerns; as a researcher perhaps even more, participating in the adventure of the mountains, deserts and forests of the living world and its inhabitants. (cited in Beaujeu-Garnier 1995, 304, 306)

Acknowledgements

I extend my thanks to Ron Cooke, Armand Frémont, Andrew Goudie, Hugh Prince, Andrew Warren and especially to Michael Wise for advice and information.

Bibliography and Sources

1. OBITUARIES AND OTHER ACCOUNTS RELATING TO JEAN DRESCH

Anon. (1948), 'Jean Dresch', *Annales de l'Université de Paris* 18, 401–8.

Anon. (1956), 'La Méditerranée Orientale et le Moyen Orient', *Geographical Journal* 122, 374–5.

Anon. (1994), 'Nécrologie: Jean Dresch', *Acta Geographica* 99, 71–2.

Baduel, P-R. (1986), 'Savoirs et pouvoirs à propos de Jean Dresch', *Revue de l'Occident Musulman et de la Méditerranée* 41–2, 9–18.

Baduel, P-R. (ed.) (1987), *Désert et Montagne au Maghreb: Hommage à Jean Dresch*. Aix-en-Provence: Edisud.

Bataillon, C. (1983), 'French geography in the 1940s', in A. Buttimer (ed.), *The Practice of Geography*. London: Longman, 119–40.

Bataillon, C. (2006), 'Six géographes en quête d'engagement: du communisme à l'aménagement du territoire', *Cybergéo* 341, 1–15.

Bataillon, C. (2009), *Géographes Génération 1930*. Rennes: Presses Universitaires de Rennes.

Beaujeu-Garnier, J. (1989), 'Remise du Grand Prix de la Société de Géographie au Professeur Jean Dresch', *Acta Geographia* 77, 74–6.

Beaujeu-Garnier, J. (1995), 'Jean Dresch et l'unité de la géographie', *Annales de Géographie* 104, 304–9.

Blache, J. (1943), 'Recherches sur l'évolution du relief dans le massif central du Grand Atlas, le Haouz et le Sous', *Revue de Géographie Alpine* 31, 429–34.

Chouiki, M. (2008), 'La géographie coloniale: engagement politique et flou identaire. Le cas marocain', in P. Singaravélou (ed.), *L'Empire des Géographes. Géographie, exploration et colonisation XIXe –XXe siècle*. Paris: Belin, 112–23.

Claval, P. (2000), '*Hérodote* and the French Left', in K. Dodds and D. Atkinson (eds), *Geopolitical Traditions: A Century of Geographical Thought*. London: Routledge, 239–67.

Clayton, D. and Bowd, G. (forthcoming), 'Geographical warfare in the tropics: Yves Lacoste and the Vietnam War', *Annals of the Association of American Geographers*.

Clout, H. (2011), 'Professorial patronage and the formation of French geographical knowledge. A biobibliographical exploration of one hundred non-metropolitan regional monographs, 1893–1969', *Cybergéo* 549, 1–36.

Coquery, M. and Coquery-Vidrovitch, C. (1980), 'Jean Dresch, un géographe au decline des empires', *Annales de Géographie* 89, 239–40.

Dollfus, O. (1984), 'Le regard attentive, le regard sélectif: Jean Dresch et Pierre Gourou. Entre monde tropical et tiers monde', *Hérodote* 33–4, 73–88.

Dollfus, O. (1994), 'Jean Dresch', *La Lettre d'Intergéo* 4, 1–2.

Douglas, I. (1986), 'The unity of geography is obvious', *Transactions, Institute of British Geographers* 11, 459–63.

Dusserre, A. (2009), 'Une étape de la mise en carte de la montagne marocaine (1937): le massif du Toubkal au 1: 20,000', *Mappemonde* 93, 1–14.

El Gharbaoui, A. (1978), 'Recherche géographique et engagement politique au Maroc', *Hérodote* 11, 87–99.

Fisher, W. B. (1954), 'La Méditerranée et le Moyen-Orient', *Geographical Journal* 120, 225–6.

Gendre, F. (1943), 'L'évolution du relief dans le massif ancien du Grand Atlas, le Haouz et le Sous d'après M. Jean Dresch', *Revue de Géographie Marocaine* 27, 38–48.

Goudie, A. S. and Wise, M. J. (1994), 'Jean Dresch, 1905–1994', *Geographical Journal* 160, 359.

Harris, C. D. (1974), 'Minutes of the Executive Committee, 29–30 May', *International Geographical Union Bulletin* 25, 30–3.

Harris, C. D. (1977), 'Minutes of the 14th General Assembly, Moscow, 1976', *International Geographical Union Bulletin* 28, 9–13.

Harris, C. D. (1996), 'On political issues', in M.-C. Robic, A.-M. Briend and M. Rössler (eds), *Géographes face au monde*. Paris: L'Harmattan, 291–304.

Joly, F. (ed.) (1979), 'L'homme et son milieu naturel au Maghreb, en hommage à Jean Dresch', *Méditerranée* 35, 1–142.

Lacoste, Y. (1972), 'Bombing the dikes: a geographer's on-the-site analysis', *The Nation*, 9 October, 298–301.

Lacoste, Y. (1973), 'An illustration of geographical warfare: bombing of the dikes on the Red River, North Vietnam', *Antipode* 5(2), 1–13 [Reprinted in R. Peet (ed.), *Radical Geography*. London: Methuen, 1978, 244–61].

Lacoste, Y. (1978a), 'Un géographe anticolonialiste', *Hérodote* 11, 3–9.

Lacoste, Y. (ed.) (1978b), 'Géographie/anticolonialisme: Jean Dresch', *Hérodote* 11, 1–159.

Mainguet, M. (1984), 'Géographie des régions arides', *Annales de Géographie* 93, 600–1.

Martonne, de E. (1943), 'Le Grand Atlas de Marrakech, d'après Jean Dresch', *Annales de Géographie* 53, 49–53.

Mombert, M. (2005), 'De crise en crise: l'enseignement de l'Allemand de 1918 à 1939', *Histoire de l'Education* 106, 71–136.

Naciri, M. (1979), 'Une géographie coloniale', *Hérodote* 13, 36–59.

Naciri, M. (1984), 'La géographie coloniale: une science appliquée à la colonisation. Perceptions et interprétations du fait colonial chez J. Célérier et G. Hardy', in J.-C. Vatin (ed.), *Connaissances du Maghreb: sciences sociales et colonisation*. Paris: CNRS, 309–43.

Prenant, A. (1995), 'In memoriam. Jean Dresch, geógrafo del Maghreb', *Eria* 38, 213–15.

Raynal, R. (1963), 'Jean Célérier, 1887–1962', *Revue de Géographie du Maroc* 3–4, 7–9.

Ruiz, A. (1997), *Présence de l'Allemagne à Bordeaux du siècle de Montaigne à la veille de la Seconde Guerre mondiale*. Bordeaux: Presses Universitaires de Bordeaux.

Spire, A. (1994), 'Jean Dresch, géographe humaniste', *L'Humanité* 7 March.

Suremain, de. M.-A. (2004), 'Faire du terrain en AOF dans les années cinquante', *Ethnologie Française* 34, 651–9.

Suret-Canale, J. (1994), 'In memoriam: Jean Dresch', *La Pensée* 299, 131–4.

Tissier, J.-L. (1996a), 'Jean Dresch', in J. Julliard and M. Winock (eds), *Dictionnaire des intellectuels français*. Paris: Seuil, 370–1.

Tissier, J.-L. (1996b), 'Julien Gracq', in J. Julliard and M. Winock (eds), *Dictionnaire des intellectuels français*. Paris: Seuil, 551–3.

Warren, A. (1984), 'Géographie des régions arides', *Geographical Journal* 150, 258–9.

The Fonds Jean Dresch has been digitized by the library of the Université de Paris VIII and contains 115 of Dresch's field notebooks, dating from 1934 to 1990, accompanied by 178 photographs and two maps. These materials are accessible at: www.bibliotheque-numerique-paris8.fr/eng/notices/104686-Fonds-Jean-Dresch.html.

2. *SELECTED WORKS BY JEAN DRESCH*

1930a 'La région prérifaine: essai morphologique', *Annales de Géographie* 39, 395–415.

1930b 'Le massif de Moulay Idriss (Maroc septentrional). Étude de géographie humaine', *Annales de Géographie* 39, 496–510.

1933a *Essai sur l'évolution du relief dans la région prérifaine*. Paris: Institut des Hautes Études Marocaines.

1933b 'Le modelé de la région prérifaine du Maroc occidental dans son rapport avec le climat', in *XIIIe Congrès International de Géographie, Paris, 1931: Comptes-Rendus*, vol. 2(1). Paris: Armand Colin, 651–8.

1935 'Les caractères généraux de la vie économique dans le Haut Atlas occidental', *Annales de Géographie* 44, 424–7.

1936 'A propos des formes de nivation dans le Grand Atlas occidental', in *XIVe Congrès International de Géographie, Varsovie, 1934: Comptes-Rendus*, vol. 2. Warsaw: Union Géographique Internationale, 224–8.

1936 (with A. Cholley and R. Clozier) *La France: métropole et colonies*. Paris: Baillière.

1938 'Les surfaces du piémont dans les Djebilet et le massif central du Grand Atlas', in *XVe Congrès International de Géographie, Amsterdam, 1938: Comptes-Rendus,* vol. 2(a). Leiden: Brill, 133–40.

1938 (with J. de Lépiney) *Guide alpin de la montagne marocaine: le massif du Toubkal.* Rabat: Office chérifien du tourisme.

1941a *Recherches sur l'évolution du relief dans le massif central du Grand Atlas, le Haouz et le Sous.* Tours: Arrault (reprinted 1984, Caen: Paradigme).

1941b *Documents sur les genres de vie de montagne dans le massif central du Grand Atlas.* Tours: Arrault.

1942–5 'La structure de l'Afrique du Nord', *L'Information Géographique* 9, 46–53.

1945 'Les migrations des populations des colonies françaises vers la Gold Coast', *Bulletin de l'Association de Géographes Français* 171–2, 84–92.

1945 'Lyautey', in C.-A. Julien and R. Delavignette (eds), *Les techniciens de la colonisation.* Paris: Presses Universitaires de France, 133–56.

1946 'Sur une géographie des investissements de capitaux. L'exemple de l'Afrique Noire', *Bulletin de l'Association de Géographes Français* 177–8, 59–64.

1947a 'Méthodes coloniales au Congo belge et en Afrique équatoriale française', *Politique Etrangère* 1, 77–89.

1947b 'Pénéplaines africaines', *Annales de Géographie* 56, 125–37.

1948a 'Villes congolaises. Étude de géographie urbaine et sociale', *Revue de Géographie Humaine et d'Ethnologie* 3, 3–24.

1948b 'La prolétarianisation des masses indigènes en Afrique du Nord', *Chemins du Monde,* 57–68.

1948 (with A. Cholley and R. Clozier) *La France et l'Union française.* Paris: Baillière.

1949 'La riziculture en Afrique occidentale', *Annales de Géographie* 58, 295–312.

1950 'Villes d'Afrique occidentale', *Cahiers d'Outre-Mer* 3, 200–30.

1951 'Sur les pédiments en Afrique méditerranéenne et tropicale', *XVIe Congrès International de Géographie, Lisbon 1949: Comptes-Rendus,* vol. 2. Lisbon: International Geographical Union, 19–28.

1952 'Les investissements en Afrique Noire', *Présence Africaine* 13, 232–41.

1952 (with M. Gigout, F. Joly, J. Le Coz and R. Raynal) *Aspects de la géomorphologie du Maroc.* Rabat: Notes et Mémoires du Service Géologique.

1953a 'Recherches sur les investissements dans l'Union française Outre-Mer', *Bulletin de l'Association de Géographes Français* 231–2, 2–13.

1953b 'Aspects nouveaux de l'Afrique du Nord', *Annales de Géographie* 62, 37–56.

1953/56 (with P. Birot) *La Méditerranée et le Moyen-Orient,* 2 vols. Paris: Presses Universitaires de France [later translated into Russian].

1955a 'Population et ressources de la Chine nouvelle', *Annales de Géographie* 64, 177–201.

1955b 'L'enseignement de la géographie et la recherche en Union soviétique', *Annales de Géographie* 64, 378–86.

1955 (with P. Birot and R. Capot-Rey) 'Recherches morphologiques dans le Sahara central', *Travaux de l'Institut de Recherches Sahariennes* 13, 13–74.

1956a 'Emmanuel de Martonne', *Bulletin de la Société Géologique de France* 6, 623–42.

1956b 'Le fait national algérien', *La Pensée*, 3–13.

1956c 'Études sur l'Afrique du Nord', in G. Chabot, R. Clozier and J. Beaujeu-Garnier (eds), *La géographie française au milieu du XXe siècle*. Paris: Baillière, 219–31.

1957a 'Dépôts de couverture en relief en Afrique occidentale française', *17th International Geographical Congress, Washington, D.C. 1952: Proceedings*. Washington: National Academy of Sciences, 323–6.

1957b 'Pédiments et glacis d'érosion, pédiplaines et inselbergs', *L'Information Géographique* 21, 183–96.

1957 (with C. Robequain) 'Progrès de la géographie dans les territoires français de l'Afrique tropicale y compris Madagascar', *17th International Geographical Congress, Washington, D.C. 1952: Proceedings*. Washington: National Academy of Sciences, 81–5.

1958 'Problèmes morphologiques des Andes centrales', *Annales de Géographie* 67, 130–51.

1960a 'Remarques sur les surfaces d'aplanissement et les reliefs résiduels en Afrique tropicale', in *XVIIIe Congrès International de Géographie,* Rio de Janeiro, *1956: Comptes-Rendus*, vol. 2(2). Rio de Janeiro: Comité National du Brésil, 213–19.

1960b 'Dépressions fermées encaissées en régions sèches, spécialement en Afrique du Nord', in *XVIIIe Congrès International de Géographie,* Rio de Janeiro, *1956: Comptes-Rendus,* vol. 2(2). Rio de Janeiro: Comité National du Brésil, 222–8.

1960c 'Rencontre franco-soviétique', *Annales de Géographie* 69, 61–4.

1961 (with P. Bout, M. Derruau and C. Peguy) 'Observations de géographie physique en Iran septentrional', *Mémoires et Documents du Centre de Documentation Cartographique et Géographique: Centre National de la Recherche Scientifique* 8, 9–101.

1963 'Développement et localisation industriels au Maghreb', in J. Dresch et al., *Industrialisation au Maghreb*. Paris: Maspero, 7–23.

1963 'Réforme agraire et sous-développement dans le monde arabe', in J. Dresch et al., *Réforme agraire au Maghreb*. Paris: Maspero, 9–20.

1966a 'Utilization and human geography of deserts', *Transactions of the Institute of British Geographers* 40, 1–10.

1966b 'Les paysages tropicaux humides; La zone aride', in A. Journaux, P. Deffontaines and M. J. Brunhes-Delamarre (eds), *Géographie générale*. Paris: Gallimard – Encyclopédie La Pléiade, 609–780.

1966 (with P. Birot) 'Pédiments et glacis dans l'ouest des Etats-Unis', *Annales de Géographie* 75, 513–52.

1966 (with P. George) 'Raoul Blanchard', *Annales de Géographie* 75, 1–15.

1968a 'L'Institut de Géographie', *Annales de l'Université de Paris* 38, 374–8.

1968b 'Cinquante ans de géographie soviétique', *Annales de Géographie* 77, 513–15.

1970 'Quelques observations dans la région de Darjeeling', *Annales de Géographie* 79, 157–65.

1972 'A propos des limites inférieures des phénomènes cryonivaux et glaciaires quaternaires aux marges méridionales et orientales du domaine méditerranéen', in *International Geography, XXII International Geographical Congress, Montreal, 1972*, vol. 1. Toronto: University of Toronto Press, 110–12.

1972 (with A. Journaux) 'Phénomènes périglaciaires en Sibérie orientale', *Bulletin de l'Association de Géographes Français* 395–6, 57–73.

1973 (with P. George) 'Le XXIIe Congrès International de Géographie, Montréal 1972', *Annales de Géographie* 82, 385–420.

1973 (with J. Besancenot and J. Tricart) 'Observations sur les processus morphogénétiques "froids" au Liban', *Revue de Géographie Physique et de Géomorphologie Dynamique* 15, 231–72.

1975 'Emmanuel de Martonne, 1873–1955', in P. George (ed.), *Les géographes français*. Paris: Comité des Travaux Historiques et Scientifiques, 35–48.

1976 'The evaluation and exploitation of the West African Sahel', *Philosophical Transactions of the Royal Society London B* 278, 537–42.

1977a 'Discours du Président, 28 juillet 1976', *International Geographical Union Bulletin* 28, 1–9.

1977b 'Crise de la géographie?', *La Pensée* 194, 19–27.

1977c 'Géographie et Sahel', *Hérodote* 6, 54–71.

1978 'Un demi-siècle de recherché géomorphologiques', *Hérodote* 12, 11–51.

1979 *Un géographe au déclin des empires*. Paris: Maspero.

1980a 'Le désert du Lout en Iran. Un désert absolut est-il infranchissable?', *Hérodote* 18, 46–56.

1980b 'Désertification et Tiers-Monde', *La Pensée* 212, 18–28.

1982 *Géographie des régions arides*. Paris: Presses Universitaires de France.

1984a 'Il n'y a pas de géographie sans drame', *Hérodote* 33–4, 33–49.

1984b 'Géographie d'hier et d'aujourd'hui', *La Pensée* 239, 9–20.

1986a 'Jean Dresch et le Maghreb: entretien', *Revue de l'Occident Musulman et de la Méditerranée* 41–2, 19–26.

1986b 'Innokententiy Petrovich Gerasimov, 1905–1985', *Annales de Géographie* 95, 233–6.

A bibliography of the published works of Jean Dresch, containing more than 300 items, is given in Baduel, P.-R. (ed.) (1987), *Désert et Montagne au Maghreb: Hommage à Jean Dresch*. Aix-en-Provence: Edisud, 27–42; and in Anon. (1987), 'Jean Dresch. Publications 1930–1986', *Revue de l'Occident Musulman et de la Méditerranée* 41–2, 27–42.

Chronology

1905	Born 30 November, in Paris
1917–23	Educated at *lycée* in Bordeaux
1923–6	Prepared in Toulouse for entry to the École Normale Supérieure (rue d'Ulm), in Paris
1926–8	Studied at the École Normale Supérieure, obtained first degree from the Sorbonne
1928–9	First visit to Morocco
1929	Completed *diplôme d'études supérieures* at the Sorbonne
1930	Passed *agrégation d'histoire et géographie* competitive examination; undertook military service, and married Andrée Berger
1931–40	Taught at the Collège musulman and the *lycée* in Rabat, Morocco
1939–40	Served in the army
1941	Awarded *doctorat d'état* at the Sorbonne
1941–2	Lectured on the geography of North Africa at the Sorbonne
1941–4	Taught at the Lycée Voltaire in Paris
1942–6	Taught geography at the University of Caen
1945	First visit to tropical Africa
1946–7	Occupied personal chair of geography at the University of Strasbourg
1946–78	Lectured at the Institut d'Études Politiques de Paris
1947–8	Taught at the École Nationale de la France d'Outre-Mer
1947–69	Lectured at the École Nationale des Langues Orientales Vivantes
1948	Appointed Professor of Geography and Colonization of North Africa at the Sorbonne
1950–78	Taught at the École Normale Supérieure (rue d'Ulm)
1953–6	Publication of *La Méditerranée et le Moyen-Orient*, 2 vols (with Pierre Birot)
1954	First visit to China
1955	First visit to the USSR
1956	Appointed Professor of Physical Geography at the Sorbonne

1960–70	Head of the Institut de Géographie at the Sorbonne; also directed the *Service de documentation et de cartographie géographique*
1966	Appointed foreign member of the Academy of Sciences of the USSR
1968–72	Member of the International Geographical Union Commission on Present Day Geomorphological Processes
1970	Moved to the Université de Paris VII
1972–6	President of the International Geographical Union
1973	Made *Lauréat de l'Académie des Sciences* in Paris
1980	Made *Lauréat d'Honneur* of the IGU at the 24th Congress in Tokyo
1982	Publication of *Géographie des régions arides*
1988	Awarded the Grand Prix de la Société de Géographie de Paris
1994	Dies 6 March, in Paris

André Cholley

1886–1968

Hugh Clout

Coming from a small village in eastern France, André Cholley was educated in the city of Lyon where he was taught by Emmanuel de Martonne. Cholley's doctoral work on the Pre-Alps of Savoy embraced Vidalian regional geography and a Davisian approach to landscape analysis. Appointed to the Sorbonne in 1927, Cholley came to adopt a process-orientated approach to geomorphology, advocated a holistic conception of geography and introduced a rudimentary 'systems approach' covering both physical and human components of the discipline. His concern for the interactive combination of causative elements was articulated in articles and in his methodological textbook. As a stimulating lecturer, inspiring mentor of graduate students, author of influential textbooks, efficient editor of academic journals, friendly yet business-like faculty dean, and long-serving director of the Institute of Geography in Paris, the contribution of André Cholley to the development of geography merits attention.

Education, Life and Work

Born in 1886 in the remote settlement of Jasney in the Haute-Saône *département*, André Cholley displayed notable promise in his primary schooling. His secondary education was in the city of Lyon, where one of his teachers was Edouard Herriot (1872–1957), pupil of Paul Vidal de la Blache (*Geographers* Vol. 12) at the École Normale Supérieure in Paris, friend of the Vidalians, and future mayor of Lyon who would rise to high office as Prime Minister on three very brief occasions. After passing his school-leaving examination with flying colours, Cholley studied at the local university where Emmanuel de Martonne (*Geographers* Vol. 12) was the dynamic professor of geography and was putting the finishing touches to the first edition of his influential *Traité de Géographie Physique* (1909). He had already obtained accommodation for his Institut de Géographie in the main building of the Faculté des Lettres, which he equipped with maps, atlases and geological specimens. He aimed to revitalize the venerable Société de Géographie de Lyon by leading fieldtrips in the local area and by organizing discussions on topical issues which were very different from the Society's round of public lectures (De Martonne 1908a; Clout 2009a). He also sought to make geographical knowledge available to the merchants, bankers and industrialists of the city, throwing his *Institut* open to visitors on 16 February 1908 (De Martonne 1908b). As well as attempting to publicize his discipline, De Martonne wanted to raise funds for equipment and to support staff. André Cholley flourished alongside this ambitious professor and wrote his *diplôme d'études supérieures* (equivalent to a research Masters thesis) about his home region, the Vôge upland. As De Martonne remarked in a letter dated 24 February 1908 to his friend Albert Demangeon (*Geographers* Vol. 11): 'Young Cholley is studying the region at the source of the Saône' (Clout 2003). Cholley also wrote his first book review in 1908 and had accompanied De Martonne on a fieldtrip in the Alps in the previous July, which happened to coincide with a visit by W. M. Davis (*Geographers* Vol. 5) and eminent geologists; a joint publication was written from this experience (De Martonne and Cholley 1908). Cholley also prepared for the fiercely competitive national examination (*concours d'agrégation d'histoire et géographie*) for prospective teachers who sought to work at the highest level in the state high school (*lycée*) system or to lecture in the university.

With his first research degree completed and the *agrégation* passed successfully, Cholley had to undertake two years of military service (1910–12), in the knowledge that a teaching post awaited him at the *lycée* in Annecy from which he could undertake doctoral research on the regional geography of *Les Préalpes de Savoie, Genevois, Bauges et leur avant-pays*. This project was interrupted by the outbreak of hostilities in the summer of 1914 and Cholley spent the duration of World War I serving in the infantry, rising to the rank of captain. It was rumoured that he spent quiet moments on the Chemin des Dames drafting one of his early publications (Gras 1975, 153). In 1919, he was demobilized and intended to return to Annecy, but before the end of the year he had moved to a more attractive post at the Lycée Ampère in Lyon. This position enabled him to continue his research and also to broaden his horizons in the city's university where geography was taught by Maurice Zimmermann and several schoolteachers, following the departure of De Martonne in 1909 to occupy a post at the Sorbonne (Clout 2011). Zimmermann was a distinguished *agrégé* but was not eligible for a professorial chair since he lacked a doctorate (Cholley 1952). While completing his thesis, Cholley gave classes in the university where he was appointed lecturer in geography in 1923. He also gave his full support to the Commission d'Études Rhodaniennes that was established in October 1923 with the backing and financial support of the City of Lyon and the *département* of the Rhône. It was 'an organ of scientific organization with the task to stimulate and coordinate research and physical studies concerning the Rhône and its tributaries, to publish them, and to furnish necessary information for understanding and managing the river, and developing regions linked to it . . . through a vast scientific enquiry' (Cholley 1924–25a, 37). Cholley served as the founding *secrétaire-général* of the Commission and edited the first issues of its journal, *Études Rhodaniennes*. On 9 December 1925 he defended his 756-page major thesis, plus a secondary thesis on La Guillotière district of Lyon, at the Sorbonne. This event attracted a large audience of scholars not only from Paris but also from the Alpine city of Grenoble. In the following year, Cholley obtained a professorial chair in Lyon and devoted his energy to teaching, writing for and editing the *Études Rhodaniennes* and serving the Commission. Despite the very real attractions and opportunities of Lyon, the lure of the Sorbonne was even more appealing and in 1927 he moved to the capital to work with De Martonne and to fill the post made vacant by the retirement of Lucien Gallois (*Geographers* Vol. 24). After a year as a senior lecturer, Cholley received a personal chair in Paris with particular responsibility for teaching the regional geography of France and devising appropriate exercises in map work to accompany those classes (George 1996, 185).

Cholley was a generation younger than De Martonne and Demangeon, the professor of human geography at the Sorbonne. After a full decade devoted to teaching, research and publication, Cholley was enlisted in September 1939 as battalion commander in the intelligence and survey section of the geographical service of the French army, a post that he held until July 1940 (Anon. 1947). During World War II, he continued to teach and administer at the Sorbonne, rising to the position of vice-dean of the Faculté des Lettres. In November 1944, he moved into the established chair of geography following the retirement of De Martonne (Demangeon having died in 1940, a little short of retirement age), and was elected dean in 1945. He would hold that position until 1953, while also serving as director of the Institut de Géographie from 1944 to 1956. Unlike De Martonne, he did not consider holding high office in the International Geographical Union but he was instrumental in energizing the Association des Géographes Français and editing two academic journals. In addition, from 1946 until his retirement ten years later, he presided over the influential jury of examiners for the new *agrégation de géographie* (freshly separated from 'aristocratic history' and no longer requiring a qualification in Latin), which assessed the quality of prospective high-school teachers and university lecturers in geography (Gras 1975, 157). By assisting geography to stand alone, he enabled the discipline to start to modernize and to shed the shackles of traditional subservience to history in French higher education (Claval 1998, 175; Deneux 2006, 105). After a dozen years of retirement, during which he continued his editorial work, André Cholley died on 29 November 1968, at the age of 82.

Scientific Ideas and Geographical Thought

Like other geographers of his time, Cholley believed in the unity of his chosen discipline, the centrality of regional study to geography, the obligation to be competent in both physical and human branches and the desirability of being conversant with the latest developments in cognate disciplines, ranging from botany and geology to economics and history. In 1942, he published his influential *Guide de l'Étudiant en Géographie* which distinguished between thematic and regional geography but proclaimed the fundamental unity of the discipline to future generations of students, most of whom would become schoolteachers. In subsequent years he pioneered what may be recognized as a rudimentary form of 'systems thinking' in which a wide range of physical and human factors interacted to produce different *combinaisons* that were expressed as regions, landscapes and ways of life and were, of course, liable to change with the passage of time (Gras 1984, 185). The second edition of the *Guide de l'Étudiant* (1951) paid less attention to the unity of geography and highlighted systems of interaction. In the words of André Meynier, this little book was a 'veritable breviary' for generations of students and teachers, and Olivier Orain recently called it 'a sort of breviary of post-Vidalian realism' (Meynier 1971, 5; Orain 2009, 91). Ever mindful of the growing separation of geography from history in French universities, Cholley declared:

> In its very essence, geography is not more historic than biological. It is both one and the other; it is our task [as educators] to determine the most suitable method to penetrate the secret of its interactions (*combinaisons*). Only to give our [young] geographers a training in history is to expose them to only one aspect of things, and to deprive them of a part of the means that will allow them to seize reality. (Cholley 1948a, 90)

By virtue of a suite of research publications in physical geography and the fact that he taught physical geography after the retirement of De Martonne, Cholley is often remembered for his work on landforms, although his expertise ranged across the discipline. His earliest work on the Vôge was a model study of a small upland region in which he traced the distinctive physical characteristics of this isolated part of eastern France, covered with tracts of timber and damp valley floors, where deposits of iron ore had given rise to a range of craft industries in the past (Cholley 1914). Using archival documents and drawing on his familiarity with the area in the field, he traced the characteristics of the traditional economy and then highlighted the impact of railway services during the second half of the nineteenth century, which served to expose the area to outside influences and to encourage migration to the growing industrial towns of Lorraine and of the Saône valley. At the time of his enquiry, agriculture, forestry and traditional crafts had dwindled, rural depopulation was entrenched and abandoned farmhouses were a common sight.

André Cholley's doctoral thesis, prepared with advice from De Martonne, displayed a similar holistic approach, with two-fifths of the text being devoted to physical themes (morphology, climate, hydrology) and the remainder to aspects of human activity (transformation of the natural environment, ways of life) both in the past and in the present (Cholley 1925a). The monograph was rather unusual since it focused not only on the Pre-Alps of Savoy but also on their low-lying foreland. De Martonne lavished praise on the work of this man whom he would recognize as his favourite pupil (Broc 2001, 98). He declared that 'no question of physical, economic or even political geography has been neglected; everything is treated in a very personal way, and very often new interpretations are advanced' (De Martonne 1926, 363). He believed that 'historians would benefit from reading the 150 pages that give a precise and vibrant picture of economic life during the 18th century' and hoped that 'this first great monograph on an Alpine region would be followed by other equally complete solidly researched studies, but perhaps with a lighter structure' (De Martonne 1926, 365, 367).

Writing from the Institut de Géographie Alpine in Grenoble, Raoul Blanchard – who was not on amicable terms with De Martonne – combined praise for Cholley's work with trenchant criticism. He declared that the thesis was 'too long, with the reader not being encouraged by its weight and density. It is too long, first of all because it considers two regions instead of one' (Blanchard 1926, 675). Cholley's consideration of rainfall and hydrology, he insisted, added nothing to work already published by his own doctoral students at Grenoble. The discussion of economic conditions in medieval times was 'a bit thin' and should have been expanded with information held in the archives of Turin (Blanchard 1926, 673). The treatment of forests and routeways was open to criticism and Cholley would have done well to compare his study areas with other parts of the Alps, since Blanchard believed that comparison of regional configurations was one of the duties of a geographer. All this paled into insignificance, however, when confronted with the attack on Cholley's analysis of landscape morphology which followed 'the method of M. De Martonne' who was heavily influenced by the ideas and approaches of W. M. Davis (Blanchard 1926, 672). Instead of identifying successive erosion surfaces, Blanchard favoured a thorough examination of glacial and hydrological processes and of the significance of underlying geological structures. In fact, such a dynamic approach was not new to Cholley, since his recent paper on the Annecy valley was subtitled 'a study in fluvio-glacial morphology' (Cholley 1923). While praising Cholley's hard work and claiming to demonstrate the esteem in which he held its author, Blanchard concluded his review of the thesis in a devastating way. 'The Pre-Alps of Savoy', he declared, 'is a book of youth, which smells of theories from the schoolroom. Now that he has become a teacher in his own right, and will acquire more flexible views derived from increasing contact with reality, I am sure that M. Cholley will give us shorter books that will be more genuinely useful' (Blanchard 1926, 677). This attack was aimed at De Martonne as much as at Cholley, and was just one expression of methodological and personal differences between the geographers of Grenoble and some of those in Paris that would last for many years (Broc 2001). However, Pierre Brunet recalled an interuniversity fieldtrip to Provence in 1954 when 'a thick fog interrupted our ascent of Mont Ventoux. And there, André Cholley and Raoul Blanchard, two geomorphologists who had held violently different views, managed to reach some agreement, which was doubtless facilitated by the fact that we could see no more than 20 metres around us!' (Letter from P. Brunet, 7 July 2010).

During his eight years as schoolteacher, lecturer and professor at Lyon, Cholley investigated the physical geography of nearby areas, such as the Beaujolais, Jura and southern Burgundy that would appear in later articles or would be presented in papers to successive International Geographical Congresses (Cholley 1929, 1932a, 1935a). For example, in 1931 he discussed erosion surfaces in the southern Jura at the 13th IGC in Paris, which was orchestrated by De Martonne, and led an official fieldtrip in his thesis area with support from Georges Chabot (Chabot and Cholley 1931; Cholley 1933). Four years later, Cholley presented a map of erosion surfaces in the Charolais and Mâconnais areas to the 14th IGC in Warsaw (Cholley 1936). In addition, he wrote about various aspects of the human geography of Lyon, including studies of the suburban development of La Guillotière, of milk supply to the city, of fruit production and marketing, and notes on economic, urban and regional themes that appeared in the earliest issues of *Études Rhodaniennes* under his editorship (Cholley 1925b, 1926, 1927a, 1927b, 1928). However, his links with the city and the journal were severed soon after he moved to Paris in 1927, with his successor André Allix serving as editor and as *secrétaire-général* of the Commission d'Études Rhodaniennes.

As the junior professorial colleague of De Martonne and Demangeon at the Sorbonne, Cholley explored the regional geography of France with his students in the lecture hall and map room, and introduced them to the landscapes of the Paris basin in the field. Already open to alternative approaches to interpreting the plateaux, valleys and scarp faces of the Ile-de-France, he came to question the explanatory merits of structure, surface and stage that had been advanced by De Martonne under the influence of Davis (Meynier 1969, 64). Cholley's paper to the 15th IGC held at Amsterdam in 1938 focused on surface features in the Paris region that had been formed, transformed and retouched under palaeoclimatic conditions very different from those of recent times (Cholley 1938). Particular attention was devoted to identifying and interpreting the occurrence on ancient erosion surfaces of 'millstone' formations (silcretes) that had originated in soils under tropical or sub-tropical conditions. Cholley argued that field observation was a necessary but insufficient approach to landscape interpretation. Careful collection of deposits needed to be complemented by analysis which would eventually be achieved in the tiny laboratory that he had installed in the Institut de Géographie (Birot 1969, 130; Claval 1998, 180). Perhaps his intellectual debt to De Martonne prevented Cholley making a more forthright declaration of his growing conversion to process-orientated, climatic geomorphology, and restrictions on movement during wartime German occupation hindered further fieldwork. Nonetheless in 1943 he published a long communication offering re-interpretations of the physique of the Paris region (Cholley 1943). This would be complemented in 1950 by a major assessment of the relative merits of structural and climatic approaches to geomorphology, which was followed by a further re-appraisal of the morphological evolution of the Paris basin (Cholley 1950, 1953).

Cholley's changing methodological position, which continued to stress accurate observation of features in the field but also recognized the importance of laboratory analysis and the interaction of diverse processes across time, was not restricted to his work in physical geography. In 1946, he published a paper with the deceptively innocuous title of 'Problems of agrarian structure and rural economy' (Cholley 1946). In this ground-breaking article, he displayed his dissatisfaction with the morphological approach to farmhouse types and rural settlements, which had been developed by Demangeon and other scholars, and figured strongly in countless regional monographs written by French geographers since the start of the twentieth century. He insisted that these apparently static features which could be classified according to various typologies were the expression of differing rural economies and should not be studied in separation from them. All these rural features reflected the interaction of countless economic, social, cultural, political and technological conditions that were subject to change through time, sometimes at a scarcely perceptible rate but also in rapid and dramatic ways. Taken together, these elements gave rise to interactive systems (*combinaisons*) that were reflected in the spatial configuration of ecosystems, enterprises, landscapes, regions and nations (Buttimer 1971, 139). However, consideration of interactive and dynamic human conditions alone fell far short of geographical reality. What was needed was a thorough grasp of physical elements in the total environment and an understanding of how these were appraised and re-appraised by successive human groups to create and exploit the necessary resource base for human survival and well-being.

Working from his re-interpretation of methodologies in physical geography, Cholley advanced the importance of what might be recognized as a rudimentary form of systems thinking while proclaiming the need to consider both physical and human elements as components of a single analytical domain. His view of the discipline was both old and new, both traditional and forward looking. It could accommodate Vidalian principles of geography as a 'science' (different from bookish history), of interaction between people and their natural environment, of areal differentiation and of ways of life (*genres de vie*). It also embraced the 'mode of production' approach that was being developed at the Sorbonne by Cholley's younger colleagues of Marxist persuasion, such as Pierre George (*Geographers* Vol. 29) and Jean Dresch (Bataillon 2006) (On Dresch, see *Geographers* this volume). As a man of the political left, Cholley was not uncomfortable working with scholars who belonged to the Communist Party and would, in due course, include their books in the series he edited. His modernizing approach to the discipline may be seen as opening the way for systems thinking that would permeate French academic geography during the final quarter of the twentieth century, carrying with it theorization, quantification, spatial planning, environmental planning and the study of globalization (Tricart 1991, 605–9).

Influence and Spread of Ideas

André Cholley has been largely overlooked in the history of academic geography, although he exercised an important influence not only in Lyon and Paris but also among *lycée* pupils and their teachers, undergraduates and doctoral students at the Sorbonne and fellow academics throughout France. Nor were his talents always recognized abroad, since he only wrote in French and his work was not translated. At Lyon, his dynamism was channelled into teaching, editing and management of the Commission d' Études Rhodaniennes. In Paris, he taught more than half the total of undergraduate geographers in France for three decades, covering regional geography, map work, geomorphology and climatology (Gras 1975, 154). Describing his first contact with 57-year-old Cholley in October 1943, Jacques Gras recalled,

> quite a tall and handsome man, but not overweight, who gave the impression of power, even of elegance; his hair, tinged with silver, was swept back from his forehead; a courteous smile lit up his face as he took care to skirt round the students sitting near the lecturer's desk, whilst enjoying the fact that he attracted such a large audience . . . His lectures were very clear and well organized. He appealed to listeners because he shared the advance of knowledge with them. He was full of new ideas, of references to work in progress; in a word, he opened perspectives on all sorts of research. He had the supreme quality of declaring when he reached the limit of his knowledge and in this way he stimulated some to become academics. His thoughts were always firmly structured [and] he formed his sentences with care, without ever raising his voice. From all this emerged a remarkable force of persuasion. (Gras 1975, 155)

As an undergraduate contemporary of Jacques Gras, Pierre Brunet confirmed that Cholley 'was a remarkable teacher as much in human geography as in physical geography. Between 1943 and 1946 there was a real struggle to obtain a seat in the amphitheatre to listen to his lectures; I remember that sometimes one had to attend the preceding lecture – of no direct relevance – simply to be sure of obtaining a place for his course. His commentaries on the 1:80,000 topographic maps were also models of their kind. His influence was essentially through the spoken word and his writings, since he was not very available to undergraduates' because of his many commitments (letter from P. Brunet, 7 July 2010). Armand Frémont remembered 'a very great professor who hypnotised the amphitheatre [in the Institut de Géographie] in rue Saint-Jacques with his geomorphology classes that were clear and illuminating' (pers comm. A. Frémont, 24 July 2010). Summarizing his experience of Cholley after almost 60 years, François Gay recalled: 'I followed his regional course for one year. He was an excellent teacher. I have, for instance, a vivid memory of his lectures on Burgundy and the wine country. His textbooks for *lycée* pupils were excellent and were much appreciated by teachers. I had a couple of fieldtrips that he led to areas to the south of Paris, and these were very interesting' (pers. comm. F.-J. Gay, 3 June 2010). Pierre Brunet described Cholley's day-long excursion, held once a year, across the valley of 'the Ru de Gally, to the west of the park of Versailles. We met one glacially cold winter's morning at the station of Les Clayes-sous-Bois, and there he delivered a preparatory talk for 45 minutes, quite indifferent to the girls who were freezing! Then followed a traverse on foot, broken by a packed lunch, leading to the railway station of Noisy-le-Roi for our return to Paris' (letter from P. Brunet, 7 July 2010). Other scholars remembered Cholley's very real skill as a field teacher, characteristically 'standing on some high point and explaining the landscape to a captivated audience' (Anon. 1969, 7).

Jacques Gras insisted that Cholley was 'naturally straightforward [and] spontaneously democratic. He behaved the same way with a humble student, a technician or secretary as he did with his colleagues, perhaps even better. He always sought to support those who were disliked by the University administration' and, for example, fixed the times of certain classes to suit trainee teachers and various part-time students who followed his courses (Gras 1975, 156). Towards the end of his career, he offered consultations to students during office hours on Sunday mornings (Bataillon 2009, 43). His profound sense of service was expressed through his commitment to the Association des Géographes Français that had been founded by De Martonne in 1920, roughly along the lines of the Association of American Geographers. After many years as secretary of the organization and publishing many research notes in its *Bulletin*, Cholley duly became its president (Anon. 1969, 7). As a still relatively young professor, he worked with Demangeon to produce a collection of teaching materials entitled *France, métropole et colonies*, which comprised a collection of photographs, maps, tables and texts examining 18 regions of France (Demangeon and Cholley, 1932). A companion volume, by Demangeon and the tropical expert Charles Robequain, covered France's colonial territories. In 1935, Cholley collaborated with De Martonne to produce a two-volume work of map interpretation exercises that covered various parts of France using the 1:80,000 *carte d'état-major* as the prime source (De Martonne and Cholley 1935). In the following year, Cholley took the lead in editing a 554-page advanced textbook for the publishing house of J.-B. Baillière. Working to the familiar title, *La France, métropole et colonies*, this volume included over 200 maps and diagrams, and over 70 photographs. Co-authors were Jean Dresch, at that stage teaching at the Rabat *lycée* in Morocco, and René Clozier, who taught at the Lycée Saint-Louis in Paris as well as lecturing at the École Normale Supérieure at Fontenay. Pierre Birot and Pierre George assisted Cholley with other textbooks in the Baillière series to accompany the 'new geography syllabus', which were used widely in France and the colonies. Ever mindful of his educational role, Cholley used his contacts with Baillière to launch a new journal for geography teachers in 1936 under the title *L'Information Géographique*. He would be the sole editor until 1945, when he assumed editorship of the *Annales de Géographie* in succession to De Martonne. Despite having acquired responsibility for the most prestigious geographical periodical in France, Cholley did not relinquish his commitments to the editorial board of the teachers' journal. In 1948 Baillière brought out a much revised version of its flagship text, re-titled *La France et l'Union Française*, by Cholley, Dresch and Clozier.

Given his devotion to his students and his colleagues, André Cholley served with distinction as vice-dean of the Faculté des Lettres at the Sorbonne, using his position to help many young men to avoid being sent to work in wartime Germany (Maigron 1994, 134). When a new dean was sought in 1945, he was elected with acclaim and bore that 'crushing, inhuman burden' in an efficient and friendly way until 1953; indeed, it 'needed his strength, authority and smiling approach . . . to accomplish that role to the satisfaction of all' (Anon. 1969, 7). Cholley was also director of the Institut de Géographie from 1944 to 1956 and continued with a heavy teaching load at a time of rising student numbers and considerable change in academic staff. In the opinion of Claude Bataillon: 'Cholley surpassed his colleagues by his gifts as a teacher, and also through his vision of geography that strove to be global' (Bataillon 2006, 6). His lectures on the rapidly changing geography of North Africa were truly fascinating (Bataillon 2009, 44). Despite all these responsibilities, he found time to encourage potential research students who sought to write theses right across the broad field of geography, not only in France but also abroad. Jacques Gras recalled that Cholley had joined him on a bicycle trip in the Marne valley in November 1945 and how, at lunch time, 'sitting on an earth bank, with a rucksack at his feet, he quietly decided how my career might develop and how I might find a place in the academic world' (Gras 1975, 156). Similarly, Pierre Pédelaborde cited Cholley's climatology lectures in 1945–6 as determining his own vocation as a climatologist (Gras 1975, 157). Despite his many commitments, Cholley was happy to advise Armand Frémont as he prepared his *diplôme d'études supérieures* on the rural geography of the Pays de Caux, which 'gave him a change from physical geography' (pers comm. A. Frémont, 24 July 2010). He stressed the importance of the concept of *combinaisons* and, in due course, enabled Frémont to publish his results in the *Annales de Géographie*. Half a century later, Frémont would declare: 'I owe a great deal to him'.

Nonetheless, encouragement from Cholley did not always involve close supervision. Pierre Brunet's *diplôme d'études supérieures* on the regional geography of the Seine valley between Montereau and Melun involved only two meetings, and

> during the eight years that I was preparing my doctoral thesis [on the rural geography of a vast area between the rivers Oise and Seine] we had only three meetings, which were not unhelpful. Faced with my discouragement at covering such a large region – that he had assigned to me – he reminded me that a principal element in a *combinaison* (in my case, large agricultural holdings) was the requirement to make useful comparisons involving other types of farm in my study area and thereby to understand how large holdings came into being and became so important. Moreover, I remain profoundly grateful to him for what I consider to be his fundamental contribution to rural geography: the notion of the *combinaison*, conceived as an holistic phenomenon . . . . His article in 1946 was my guide and I believe that I illustrated it well in my own work. (letter from P. Brunet, 7 July 2010)

Many other French academics, from Pierre George whose doctorate was completed in 1935 to Jacques Gras who finished in 1963, found their inspiration in André Cholley, welcomed him as their *patron*, and dedicated their theses to him (Clout 2009b). For a man who is now largely forgotten, his influence on such scholars as Jacqueline Beaujeu-Garnier (*Geographers* Vol. 28), Pierre Brunet and Michel Philipponneau researching in France, André Blanc in Yugoslavia, Philippe Pinchemel (*Geographers* Vol. 29) in England, Xavier de Planhol in Turkey and many others who became major figures in French geography, was truly remarkable.

Given his multitude of commitments at the Sorbonne, to which he added occasional lecturing at the Institut Géographique National, the college for Hautes Études Commerciales and the École Normale at Sèvres, it is not surprising that Cholley was not more productive (Gras 1975, 154). As Pierre Brunet insisted, Cholley's articles and spoken word formed his real legacy. A dozen of his geomorphological writings, plus his 'Problems of agrarian structure and rural economy' article, were reprinted as a sort of retirement *festschrift* under the title *Recherches morphologiques* (Cholley 1957a; Le Lannou 1968, 360). Cholley failed to write a projected textbook on France that would have drawn on his long experience at the Sorbonne. Instead he encouraged his Parisian colleagues and friends from Lyon to produce substantial university-level texts under his editorship (Birot 1969, 130). The first of these was the 'Orbis' series for the Presses Universitaires de France, which included regional themes by such authors as Pierre George, Jean Dresch and Jacqueline Beaujeu-Garnier, and systematic titles by academics including Pierre Birot, André Guilcher and Jean Tricart (Table 1). This was complemented by the 'Géographie économique et sociale' series published by Éditions Génin, which comprised various titles on agriculture, energy, industry and population, with his editorship concluding in 1968 with *Régions, nations, grands espaces* by Paul Claval. As a mark of Cholley's wide experience and the high regard in which he was held by his colleagues, he was chosen to write the introductory essay in a volume that presented the achievements of French geography to the global academic audience (Cholley 1957b).

Table 1: The 'Orbis' and 'Géographie économique et sociale' series, edited by André Cholley

Orbis (Presses Universitaires de France)	*Géographie économique et sociale* (Génin)*
URSS, Haute-Asie, Iran, P. George (1947)	*Géographie Agraire*, D. Faucher (1949)
La Méditerranée et le Moyen Orient, 2 vols, P. Birot and J. Dresch (1953, 1956)	*Géographie de l'Energie*, P. George (1950)
L'Europe Centrale, 2 vols, P. George and J. Tricart (1954)	*Géographie des Textiles*, A. Allix and A. Gibert (1957)
Morphologie Littorale et Sous-marine, A. Guilcher (1954)	*Géographie de la Population*, 2 vols, J. Beaujeu-Garnier (1958)
Les Méthodes de la Morphologie, P. Birot (1955)	*L'Économie des Transports Terrestres*, R. Clozier (1963)
Morphologie Structurale, P. Birot (1958)	*La Circulation Maritime*, A. Vigarié (1968)
L'Europe du Nord et du Nord-Ouest, 3 vols, G. Chabot, A. Guilcher and J. Beaujeu-Garnier (1958, 1958, 1963)	*Régions, Nations, Grands Espaces*, P. Claval (1968)
Géomorphologie des Régions Froides, J. Tricart (1963)	[*Géographie du Transport Aérien*, M. Dacharry (1981)]

* An earlier list shown in Faucher, *Géographie Agraire* (1949) proposed titles by Hassan Awad, Pierre Birot, Georges Chabot, Jean Dresch, Maurice Le Lannou, Charles Robequain, Francis Ruellan, and L. Dudley Stamp, but none of these books materialized in the series.

Conclusion

The career of André Cholley covered a wide span across the twentieth century, with his first research being completed before the outbreak of World War I and his death coming soon after the social revolution of 1968 that would transform the Sorbonne where he had taught for almost 30 years. His scholarship began in the vibrant heyday of Vidalian geography and then embraced many years working alongside De Martonne and Demangeon when French geography was widely regarded as being the best in the world. At the conclusion of World War II, Cholley assumed a heavy administrative burden in addition to his teaching and editorial commitments, hence his personal productivity waned. Nonetheless, it was at this time that he supported many doctoral students, advanced new ideas in geomorphology and presented his path-breaking notion of *combinaisons*, which may be seen as a prototype for systems thinking. In the words of Claude Bataillon: 'he leaves the memory of a very fine teacher, a straight talker, who behaved as a man of the people (*d'allure plébéienne*), which was not usual at the Sorbonne at that time' (Bataillon 2006, 5). As a dynamic lecturer, mentor, administrator and editor, André Cholley deserves to be recognized as a leader of academic geography in twentieth-century France.

Acknowledgements

For advice and information, I extend my thanks to Pierre Brunet, Armand Frémont, François Gay, Hugh Prince and John Catt.

Bibliography and Sources

1. OBITUARIES AND OTHER ACCOUNTS RELATING TO ANDRÉ CHOLLEY

Anon. (1947), 'André Cholley', *Annales de l'Université de Paris* 17, 227–9.

Anon. (1969), 'André Cholley', *L'Information Géographique* 33, 7.

Bataillon, C. (2006), 'Six géographes en quête d'engagement: du communisme à l'aménagement du territoire', *Cybergéo* 341.

Bataillon, C. (2009), *Géographes génération 1930*. Paris: L'Harmattan.

Birot, P. (1969), 'André Cholley', *Annales de Géographie* 78, 129–30.

Blanchard, R. (1926), 'Les Préalpes de Savoie (Genevois, Bauges) et leur avant-pays', *Revue de Géographie Alpine* 14, 671–7.

Broc, N. (2001), 'École de Grenoble contre École de Paris: les Alpes enjeu scientifique', *Revue de Géographie Alpine* 89, 95–105.

Buttimer, A. (1971), *Society and Milieu in the French Geographic Tradition*. Chicago: Rand McNally.

Claval, P. (1998), *Histoire de la géographie en France de 1870 à nos jours*. Paris: Nathan.

Clout, H. (2003), 'In the shadow of Vidal de la Blache', *Journal of Historical Geography* 29, 336–55.

Clout, H. (2009a), 'Popularising geography in France's second city: The rôle of the Société de Géographie de Lyon, 1873–1968', *Cybergéo* 449, 1–20

Clout, H. (2009b), *Patronage and Production of Geographical Knowledge in France: The Testimony of the First Hundred Regional Monographs, 1905–1966*. London: RGS/IBG, Historical Geography Research Series, No. 41.

Clout, H. (2011), 'Geographical pioneers in Lyon, 1874–1929', *Géocarrefour* 86, 161–71.

Deneux, J.-F. (2006), *Histoire de la pensée géographique*. Paris: Belin.

George, P. (1996), 'La géographie régionale à travers l'enseignement d'André Cholley, 1927–56', in P. Claval and A.-L. Sanguin (eds), *La géographie française à l'époque classique*. Paris: L'Harmattan, 185–8.

Gras, J. (1975), 'André Cholley', in P. George (ed.), *Les géographes français*. Paris: Comité des Travaux Historiques et Scientifiques, 153–71.

Gras, J. (1984), 'André Cholley, 1886–1968', in P. Pinchemel, M.-C. Robic and J.-L. Tissier (eds), *Deux siècles de géographie française: choix de textes*. Paris: Comité des Travaux Historiques et Scientifiques, 184–5.

Le Lannou, M. (1968), 'André Cholley, toujours présent', *Revue de Géographie de Lyon* 43, 359–60.

Maigron, G. (1994), 'Resistance et collaboration dans l'Université de Paris sous l'Occupation', in A. Guesclin (ed.), *Les Facs sous Vichy: étudiants, universitaires et universités de France pendant la seconde guerre mondiale.* Clermont-Ferrand: Presses Université Blaise-Pascal, 133–42.

Marconis, R. (2000), *Introduction à la géographie.* Paris: Armand Colin.

Martonne, E. De. (1908a), 'Le développement et l'avenir de la géographie', *Bulletin de la Société de Géographie de Lyon* 1, 1–11.

Martonne, E. De. (1908b), 'Visite à l'Institut de Géographie', *Bulletin de la Société de Géographie de Lyon* 1, 124–5.

Martonne, E. De. (1926), 'Les Préalpes de Savoie, d'après A. Cholley', *Annales de Géographie* 36, 363–7.

Meynier, A. (1969), *Histoire de la pensée géographique en France.* Paris: Presses Universitaires de France.

Meynier, A. (1971), *Le guide de l'étudiant en géographie.* Paris: Presses Universitaires de France.

Orain, O. (2009), *De plein-pied dans le monde.* Paris: L'Harmattan.

Tricart, J. (1991), 'Cent ans de géomorphologie dans les *Annales de Géographie*', *Annales de Géographie* 100, 578–616.

2. SELECTED WORKS BY ANDRÉ CHOLLEY

1908 (with E. De Martonne) 'Excursion géographique dans les Alpes du Dauphiné, Vercors et Oisans', *Bulletin de la Société de Géographie de Lyon* 1, 201–41.

1914 'La Vôge', *Annales de Géographie* 23, 219–35.

1923 'Les débouchés de la Vallée d'Annecy: étude de morphologie fluvioglaciaire', *Annales de Géographie* 32, 193–209.

1924–25a 'Note sur l'Institut des Études Rhodaniennes', *Bulletin de la Société de Géographie de Lyon*, 37–9.

1924–25b 'Mulhouse, son rôle industriel en Alsace et dans la France de l'Est', *Bulletin de la Société de Géographie de Lyon*, 14–23.

1925a *Les Préalpes de Savoie (Genevois, Bauges) et leur avant-pays.* Paris: Armand Colin.

1925b 'Essai d'histoire topographique de La Guillotière et de la périphérie orientale de Lyon', in A. Kleinclausz (ed.), *Lyon, des origines à nos jours.* Lyon: Masson, 3–28.

1926 'Chronique rhodanienne', *Études Rhodaniennes* 2, 181–6, 195–200.

1927a 'Quelques remarques sur la culture et le commerce des fruits dans la banlieue de Lyon', *Études Rhodaniennes* 3, 83–107.

1927b 'Chroniques', *Études Rhodaniennes* 3, 112–52.

1927–8 'L'Atlas photographique du Rhône', *Bulletin de la Société de Géographie de Lyon*, 41–3.

1928 'Culture et commerce des fruits dans la banlieue lyonnaise', *Annales de Géographie* 37, 355–9.

1929 'Notes de géographie beaujolaise', *Annales de Géographie* 38, 26–46.

1930 'L'Atlas photographique du Rhône', in *XIIth International Geographical Congress Cambridge, 1928*. Cambridge: Cambridge University Press, 118–21.

1930 (with G. Chabot) 'Notes de morphologie karstique du Poljé de Lika au Popovo', *Annales de Géographie* 39, 270–85.

1931 (with G. Chabot) *Jura-Préalpes de Savoie: Excursion A1*. Paris: Armand Colin.

1932a 'Études morphologiques sur le Jura méridional et l'Ile Crémieu: influence de la tectonique hercynienne, surface tertiaires polygéniques', *Annales de Géographie* 41, 561–82.

1932b 'L'Atlas photographique du Rhône', in *XIIIe Congrès Géographique International, Paris: Comptes-Rendus*, vol. I. Paris: Armand Colin, 164.

1932 (with A. Demangeon) *France, métropole et colonies*. Paris: Librairie de l'Enseignement.

1933 'Recherches sur les surfaces d'aplanissement tertiaires dans le Jura méridional', in *XIIIe Congrès Géographique International, Paris: Comptes-Rendus*, vol. II(1). Paris: Armand Colin, 504–16.

1934 'Essai d'une carte de la répresentation de l'habitat rural', in *XIIIe Congrès Géographique International, Paris: Comptes-Rendus*, vol. III. Paris: Armand Colin, 122–33.

1935 'Remarques sur la morphologie de la Bourgogne méridionale', *Annales de Géographie* 44, 574–94.

1935 (with E. de Martonne) *La France: interprétation géographique de la carte d'état-major à 1: 80,000*. Paris: Armand Colin.

1936 'Carte des surfaces d'aplanissement du Charolais et du Mâconnais', in *XIVe Congrès Géographique International, Varsovie: Comptes-Rendus*, vol. II. Warsaw: International Geographical Union, 456–64.

1936 (with R. Clozier and J. Dresch) *La France: métropole et colonies*. Paris: Baillière.

1937 'Etude d'une représentation cartographique de l'habitat rural dans deux régions françaises: le Valois et la Haute Tarentaise', in *XIVe Congrès Géographique International, Varsovie: Comptes-Rendus*, vol. III. Warsaw: Union Géographique Internationale, 431–2.

1938 'Surfaces meuliérisées de la région parisienne', in *XVe Congrès Géographique International, Amsterdam: Comptes-Rendus*, vol. II. Leiden: Brill, 66–82.

1942a *Guide de l'Étudiant en Géographie*. Paris: Presses Universitaires de France.

1942b 'Problèmes du bassin parisien', *L'Information Géographique* 8, 23–32.

1943 'Recherches sur les surface d'érosion et la morphologie de la région parisienne', *Annales de Geographie* 52, 1–19, 81–97, 161–89.

1946 'Problèmes de structure agraire et d'économie rurale', *Annales de Géographie* 55, 81–101.

1948a 'Remarques sur quelques points de vue géographiques', *L'Information Géographique* 12, 85–90, 127–33.

1948b 'Géographie et sociologie', *Cahiers Internationaux de Sociologie* 5, 3–20.

1948 (with R. Clozier and J. Dresch) *La France et l'Union Française*. Paris: Baillière.

1950 'Morphologie structurale et morphologie climatique', *Annales de Géographie* 59, 321–35.

1951 *Guide de l'Étudiant en Géographie*, Second edition. Paris: Presses Universitaires de France.

1952 'Maurice Zimmermann, 1870–1950', *Annales de Géographie* 61, 241–4.

1953a 'Quelques aspects nouveaux sur l'évolution morphologique du basin de Paris', *Annales de Géographie* 62, 4–17, 92–107.

1953b 'La géographie en France', *Annales de l'Université de Paris* 23, 3–13.

1956 'Emmanuel de Martonne', *Annales de Géographie* 65, 1–14.

1956 (with others) *Atlas des formes de relief*. Paris: Institut Géographique National.

1957a *Recherches morphologiques*. Paris: Armand Colin.

1957b 'Tendances et organisation de la géographie en France', in G. Chabot, R. Clozier and J. Beaujeu-Garnier (eds), *La géographie française au milieu du XXe siècle*. Paris: Baillière, 13–25.

Chronology

1886	Born at Jasney, Haute-Saône, eastern France
1898–1906	Received secondary education in Lyon
1906–10	Studied history and geography at the University of Lyon
1910	Passed *agrégation d'histoire et géographie*
1910–12	Conscripted for military service
1912–19	Held teaching post at the *lycée* in Annecy
1914–19	Underwent military service in the artillery for the duration of World War I
1919–23	Taught at the Lycée Ampère in Lyon
1923–6	Lectured in geography, University of Lyon
1925	Awarded state doctorate at the Sorbonne
1926	Promoted to the chair of geography, University of Lyon
1927	Appointed senior lecturer in geography at the Sorbonne

1928	Promoted to a personal chair of geography at the Sorbonne
1936	Founded *L'Information Géographique*
1939–40	Served in the geographical service of the French army
1942	Published *Guide de l'Étudiant en Géographie*
1944	Translated into the established chair of geography at the Sorbonne
1944–56	Directed the Institut de Géographie at the Sorbonne
1945	Assumed editorial responsibilities for the *Annales de Géographie*
1945–53	Served as Dean of the Faculté des Lettres of the Sorbonne
1951	Published second edition of *Guide de l'Étudiant en géographie*
1956	Retired from the chair of geography at the Sorbonne
1968	Dies in Paris on 29 November, aged 82

Daniel Faucher

1882–1970

Hugh Clout

Daniel Faucher came from humble origins and was a latecomer to higher education. He was trained in the Institut de Géographie Alpine in Grenoble, rather than at the Sorbonne or the prestigious Ecole Normale Supérieure in Paris, and worked in a provincial teacher-training college for over a decade. His doctoral thesis, written with encouragement from Raoul Blanchard, was a comprehensive regional monograph; Faucher was an all-rounder in both teaching and research, but his preference was for rural geography. Teaching at the University of Toulouse for a quarter of a century, he pioneered the promotion of geographical research in south-western France, and was founder of that region's geographical journal which he edited for over three decades. He was a long-serving dean of faculty and initiated many welfare projects for students and refugees, including the first student residence in Toulouse. His holistic vision of geography and his belief that knowledge of the past helped understand the present and might be of value in shaping the future were inseparable from his commitment to the cause of regional development and agricultural modernization. Two books on the Rhône valley, completed four decades apart, were complemented by important texts on rural geography and by numerous articles on many aspects of the geography of south-western France.

Education, Life and Work

Daniel Faucher was born in the small town of Romans in the lower Isère valley of southern France on 3 February 1882 (Taillefer 1970, 1975). His father was a craftsman who specialized in cabinet making, but his grandparents were farmers (Faucher 1962, i). Daniel excelled at school and in 1898 was admitted to the teacher-training college at Valence, 15 km from his home town. First as a pupil-teacher and from 1903 as a village schoolmaster, he taught at Loriol-sur-Drôme, 20 km south of Valence. There he met Jeanne Caillet, his female counterpart, and the two young teachers married in 1907. Jeanne encouraged her husband to enhance his qualifications and become an instructor in a teacher-training college. After passing the necessary examinations, Daniel joined the staff of his old college in Valence, where he taught history and geography to intending school teachers (Robic 1984). With his well-constructed lessons and friendly manner, Faucher was liked and respected by his pupils, and made an especially favourable impression on Charles Petit-Dutaillis (1868–1947), rector of the *académie* of Grenoble, during a school inspection (Blanchard 1963, 162; Papy 1971, 385). It was clear that he was a talented young man whose fascination with the economic history of his home area at the time of the French Revolution could be satisfied by studying at the local university in Grenoble. Petit-Dutaillis introduced him first to Gaston Roupnel (1871–1946), professor of local history, and then to Raoul Blanchard (1877–1965), who had been appointed to teach geography at Grenoble in 1906. Roupnel's specialism was rural society and this meeting may have been influential on the young man's career. Although Faucher's initial wish had been to focus on history, he quickly became enthused by Blanchard's geography. In any case, history and geography were studied together both in high schools (*lycées*) and universities throughout France at this time. After some coaching in Latin, Faucher passed his first degree, while still instructing at Valence. Blanchard encouraged him to begin research on the geography of the Valence area, which yielded a substantial article that appeared in 1914 in the prestigious *Annales de Géographie*, edited by Paul Vidal de La Blache (1845–1917) (*Geographers* Vol. 12) and his associates in Paris (Faucher 1914).

Aged 28 at the outbreak of World War I, Faucher was conscripted into the army and became an artillery officer, serving first in the Vosges mountains and then on the Eastern Front in Macedonia, where he gathered sufficient material for two short articles that were also accepted by the *Annales de Géographie* (Faucher 1918a, 1919). His military activities were rewarded by membership of the Légion d'Honneur. He had contracted malaria in the Balkans and required a period of recuperation before returning to his duties in Valence, indeed he would never be entirely free of the disease. These months of recovery were far from wasted, since he drafted a series of articles on the area around Valence, which appeared in the *Revue de Géographie Alpine*, edited by Blanchard, and in other local journals (Faucher 1918b, 1920). In 1919, he started to conceptualize his doctoral research that would focus on the plains and basins aligned along the Rhône valley between Bas-Dauphiné and Provence. Using Valence as his base, he undertook extensive field enquiries over the next five years and consulted documents in various archives. As was the custom of the time, regional monographs not only analysed physical conditions and contemporary aspects of socio-economic geography but also investigated the historical background to the present scene. By 1926, Faucher had completed his doctoral research and was invited to teach geography at the University of Toulouse, even before his work had been examined.

At the age of 44, he accepted this offer and succeeded Paul Dognon (1857–1931), who had given lectures in geography while being 'more of an historian' by training and preference 'than a geographer, if truth be told' (Anon. 1966, 246). Paul Dognon had studied at the prestigious Ecole Normale Supérieure in Paris, and taught history and geography at the University of Toulouse. His daughter, Suzanne, was the assistant of geographer Henri Baulig (*Geographers* Vol. 4) at the University of Strasbourg, and married his historian colleague, Lucien Febvre (*Geographers* Vol. 23).

In the following year, Faucher defended his research before a jury at the University of Grenoble, was duly awarded his doctorate, and moved into the chair of geography at Toulouse where he would remain until retirement in 1952 at the age of 70. These decades of teaching the geography of the Pyrenees and the South-West of

France enjoyed the 'discrete support' of the Société de Géographie de Toulouse, which had been founded in 1882, the year of Faucher's birth (P.-Y. Péchoux, pers. comm. 24 October 2010). The new professor acknowledged that he had no previous contact with Toulouse and was not really aware of the challenges that lay ahead (Anon. 1966, 245). He recalled: 'As I prepared to leave my native Dauphiné, I received the most cordial encouragements. My master, Raoul Blanchard, declared that I was being sent to a region where I would have much to do, since south-western France seemed to him to be a kind of unknown territory (*une sorte de terra incognita*)' geographically speaking (Anon. 1966, 245). With the wisdom of hindsight and memories of obstacles surmounted, Faucher added: 'What better could I have asked for?'

He soon discovered that provision of atlases, maps and geographical books was minimal in the University, and that no specific space was allocated to geography in the old Faculté des Lettres. The number of students who attended lectures in geography, to complement their courses in history, was small, with Faucher's first cohorts (counting all years of study together) totalling fewer than 40 scholars. They had been exposed to a very traditional, fact-laden form of geography from Dognon. Looking back, 40 years later, Faucher remarked how 'this first group of students was not very familiar with the kind of geography that I had learned and that I was going to teach' (Anon. 1966, 246). Seen by some as 'a special envoy from Raoul Blanchard', Faucher determined to practise the 'new geography' that he had discovered at Grenoble and that Blanchard had received (admittedly in an earlier form) direct from the lips of Vidal. This 'new geography' was conceived as a science in its own right (rather than as a mere ancillary to history) and embraced physical elements, map work, socio-economic aspects, regional study and fieldwork. Following the methods of Blanchard, Faucher introduced a programme of visits for his students in Toulouse and in its surrounding countryside, demonstrating the belief that geography was best learned through direct observation and discussion on the ground rather than through reading. Within a short time, these local excursions were being held every fortnight, with one or two visits being arranged each year to other parts of France or across the Pyrenees to Spain (Anon. 1966, 246). Faucher recalled how 'these early students rapidly adopted me and almost all remained my friends. I knew that employing a friendly approach generated friendship in return and, having received the benefit of friendship that my teacher in Grenoble so generously showed toward me, I wanted to give my students something of the friendship that I myself had received' (Anon. 1966, 246). Sharing in their jokes and student songs, Faucher adopted a much more familiar approach than was usual at the University of Bordeaux or in other centres of higher education in France at this time (Papy 1971, 386).

Behind this amiable exterior, the new professor of geography had a clear intellectual agenda that he was determined to pursue. First, he managed to obtain a few rooms on the second floor of the Faculté des Lettres to house a 'geographical laboratory' (following Blanchard's model in Grenoble), his office and a modest library for which special funds had to be obtained to purchase necessary books, atlases, maps and journals (Anon. 1966, 247). These rooms soon proved inadequate and Faucher convinced two faculty deans, the historian Félix Durrbach (1859–1931) and the Germanist Joseph-Emile Dresch (1871–1958) that new facilities should be set up for geography. Vacant ground on the rue de l'Université (later renamed the rue Albert-Lautmann) would eventually provide an appropriate site for a new building. After three years of work, these premises were completed in June 1939 and were occupied in the subsequent academic session; however lectures continued to be given in the main buildings of the Faculté des Lettres.

Second, Faucher mustered support in the University and the city to enable him to launch a specialist periodical, the *Revue Géographique des Pyrénées et du Sud-Ouest*, to report research undertaken by himself, by his advanced students and by local *lycée* teachers, and to carry book reviews, bibliographies and summaries of geographical information about south-western France and Iberia. Modelled on Blanchard's *Revue de Géographie Alpine*, this new periodical saw its first issue in 1930 and would be edited by Faucher (latterly in collaboration) until 1964. His faithful secretary, Madame Jeanne Reboul, provided invaluable support over many years. Unlike some French geographical journals, this periodical continued to be published in the most difficult times during World War II, albeit with unavoidable delays. From 1953, it would be co-edited by colleagues from the University of Bordeaux, and subsequently also from the new University of Pau (Faucher 1957, 200). In 1998, its title would be changed to *Sud-Ouest Européen*. Through delivery of radio talks for school teachers and ordinary citizens, involvement with many local societies, and especially through training many cohorts of students (some of whom took the demanding *agrégation* examination preparing them to operate at a high level), Faucher promoted the cause of geography across south-western France, ranging from Biarritz and Bordeaux to Périgueux and Limoges. He increased the international visibility of scholarship at Toulouse through lectures delivered in Spain, Portugal and Poland, and by his presentation of research findings at the International Geographical Congresses held in Warsaw in 1934 and at Amsterdam in 1938 (Faucher 1936, 1938). He took advantage of these occasions to widen academic contacts and to build up a collection of photographs for teaching purposes. Many of these have been conserved at the Institut de Géographie in Toulouse. Among the foreign scholars he received at this time were Wanda Rewienska and Sophie Holub Pacowicz from Poland, Ivan Batakliev from Bulgaria and Elio Migliorini from Italy (P.-Y. Péchoux, pers. comm. 24 October 2010).

Third, Faucher operated single-handed during his early years at Toulouse, like many other professors of geography in French provincial universities between the wars. However, he obtained some teaching support from local *lycée* teachers, such as Georges Jorré (1899–1956) and Lucien Goron (1886–1954), but not until 1942 was François Taillefer (1917–2006) appointed as his assistant (Péchoux 2006). Starting in 1948, local teacher Jean Sermet (1907–2003) was employed on an annual basis as an adjunct lecturer (Péchoux 2003). Additional help was provided by colleagues in other departments of the University, notably by the botanist Henri Gaussen (1891–1981) who was a specialist in vegetation mapping and taught biogeography, with special reference to forestry, in the classroom and in the field (Claval 1996, 22). Faucher reinforced his position in the University and the city by promoting services to receive and accommodate students and academics fleeing persecution during the Spanish Civil War. He strove to reinforce these welfare functions in the early 1940s to help students who migrated southward from the zone of German occupation to the so-called 'free zone'. He was instrumental in establishing the first canteens and residences for students in Toulouse, as well as supporting Resistance activities throughout World War II. With this array of credentials, he was first appointed and then elected dean of the Faculté des Lettres in 1944; he would be re-elected on two subsequent occasions. In this capacity, he worked tirelessly to acquire additional funds, academic staff and premises to cope with growing numbers of students in the range of departments in his charge (Taillefer 1970, 215). He also maintained a punishing routine of teaching, editing and writing, despite the prolonged ill-health of his wife, whose heart condition made her a semi-invalid for many years.

At this period of intense activity, friends and colleagues celebrated Faucher's 65th birthday and his 2 decades of teaching at Toulouse by contributing to a 2-volume work entitled *France méridionale et pays ibériques* (Anon. 1948–9). The 57 contributors included his 'master' Raoul Blanchard, Grenoblois colleagues (Ernest Bénévent, Paul Veyret), his own disciples (François Taillefer, Michel Chevalier, Lucien Goron, Jean Coppolani,

Simone Henry), and academic colleagues in other disciplines at Toulouse (Henri Gaussen, historian Philippe Wolff) and geographers from Iberia (e.g. Orlando Ribeiro: *Geographers* this volume).

In 1949, Faucher brought out his *Géographie agraire*, in the 'Géographie économique et sociale' series edited by Sorbonne geographer André Cholley (1886–1968), which was a much enlarged version of a small text that assembled lecture material delivered in Iberia in the mid-1930s (Faucher 1935, 1940, 1949) (on Cholley, see *Geographers* this volume). His sense of achievement at this phase of his career was darkened by the death of his wife in late September 1950. However, his interwar project for a comprehensive regional geography of France was finally realized in a two-volume work, entitled *La France, géographie, tourisme*, which was published by Hachette (Faucher 1952, 1953). This was aimed at a wide readership and reflected a novel formula, combining well-illustrated but nonetheless scholarly and fully referenced essays on the geography of each region with complementary, interwoven pieces on natural and cultural features that were appreciated by visitors. Academic geographers, in the main, were responsible for the scholarly essays, with material on tourism originating from a different team. Most geographical authors were based in the regions that they wrote about, with Faucher enlisting the help of a group of colleagues whose doctoral work had received the patronage of Raoul Blanchard and who occupied university chairs in southern and central France. As André Allix and Jean Robert – both protégés of Blanchard – remarked, the edited collection reflected 'an old comradeship among contributors who shared common approaches and very perceptible intellectual affinities that gave the work a welcome coherence' (Allix and Robert 1954, 155). Faucher also called on schoolmasters who taught in Toulouse, and on a few more distant colleagues without Grenoblois associations for coverage of eastern, northern and western France. As well as writing about the geographical diversity of France, Faucher contributed essays on the Rhône corridor, the Aquitaine basin, the Pyrenees and parts of the Mediterranean south. The absence from the collection of contributions from geographers trained at the Sorbonne in Paris was striking, with coverage of the capital city being entrusted to Pierre Lavedan, an art historian with expertise in the development of urban form. As reviewers remarked, 'this was proof of the vitality and tenacity of geographical work in the French provinces' (Allix and Robert 1954, 156). *La France, géographie, tourisme* sold well and went into three editions.

In 1952, Daniel Faucher took official retirement from the University of Toulouse, relinquishing his deanship and allowing the chair of geography to pass to his disciple and colleague, François Taillefer. Nonetheless, Faucher continued to live in Toulouse and gave his classes on rural geography for several years into retirement. Paul Claval recalled that at this time he 'remained very present' in the intellectual life of the city and was busy putting the final touches to *Le Paysan et la machine*, which appeared in print in 1954 (Claval 1996, 21). However, Claval attended only two of Faucher's classes on rural geography, which were 'an admirable spectacle with precise diction, but he had already written about everything that he taught'. Hence, Claval confessed, 'I found it more expedient to go to his books and articles' in the library 'rather than sit in the classroom' (Claval 1996, 21). Despite his age, Faucher continued to write and to present papers at conferences. His position as *doyen* of French rural geographers was acknowledged in 1951 when he analysed recent changes in ways of life in the French countryside for the Centre d'Etudes Sociologiques, and again in September 1957 when he delivered the inaugural lecture at the first international conference on the agrarian geography and history of Europe that was held in Nancy (Faucher 1953). Soon afterward, he joined historian Georges Duby and ethnologist Isaac Chiva to found the interdisciplinary journal *Etudes Rurales*, whose first number contained Faucher's seminal study of triennial crop rotations in France (Faucher 1961). He would remain co-editor of that periodical for the rest of his life. His sustained commitment to rural geography would place him on the jury of examiners for Roger Brunet's doctoral thesis on *Les campagnes toulousaines* that was held on 13 December 1965.

In 1962, Faucher's friends and colleagues celebrated his 80th birthday by assembling a collection of his published and unpublished papers, together with his specially written methodological introduction, under the title *La vie rurale vue par un géographe* (Faucher 1962). This volume charted the spectrum of his research, ranging from work on the origins of agriculture, through various historic 'revolutions' that met with varying degrees of success, to the impact of mechanization, and the challenge of increasing food supply at a time of accelerating population growth in the world, when famine still threatened the poorest countries. Recalling fieldtrips in the early 1960s, Pierre-Yves Péchoux remembers 'a tall, elderly gentleman, who was attentive and contributed clearly to discussion from time to time, but was separated by age from the geographical leaders of the time' (P.-Y. Péchoux, pers. comm. 24 October 2010). When Faucher participated in meetings of the interdisciplinary 'Association Marc Bloch', which invited visiting speakers to Toulouse, he 'appeared rather laconic, expressing general ideas rather than analyses based on observed facts' (P.-Y. Péchoux, pers. comm. 24 October 2010). In 1966, a marble plaque was installed outside the renamed 'Institut de Géographie Daniel Faucher' and thanks to the efforts of Péchoux this item would be re-erected on the wall of a new building in 1969, following the relocation of the department to its suburban site at Le Mirail. Péchoux recalled that this operation was not easy to accomplish, since the plaque had been forgotten and its reinstallation had to overcome 'a certain indifference' on the part of some geographical colleagues.

At this time, Faucher was working on a completely new book dealing with *L'homme et le Rhône* for the 'Géographie humaine' series edited by Pierre Deffontaines for the Gallimard publishing house (Faucher 1968). Exploring, in turn, the physical geography of the river and its surrounding region, the history of its exploitation, and the contemporary challenges of navigation, power generation, agricultural modernization and industrial development, Faucher established a similar structure to that of his thesis; he looked, however, to the future as well as to the past, and discussed a larger area, including the city of Lyons. Paul Veyret, declared that this was the story of 'a wild, disciplined, and mastered river' (*le Rhône sauvage, discipline, asservi*) (Veyret 1969, 212). Pierre George (*Geographers* Vol. 29), whose doctorate had focused on the lower Rhône, praised Faucher's skills as 'an historian, an economist, but always as a geographer', and concluded that *L'homme et le Rhône* was 'a great book, which only an authentic "man of the Rhône" (*rhodanien*) could write: a poetic example of historical geography, which is also prophetic, whilst remaining intensely contemporary' (George 1935; 1969, 720–1). Faucher also worked on a comparative study of rural life across southern France, but this book was not completed (Taillefer 1975, 182). Finally, he devoted his energies to a labour of love as he edited for publication a massive typescript on the rural geography of the area east of Toulouse. This research had been undertaken by his friend and colleague Jorré, who died in 1956 before his doctorate could be completed and defended. Faucher focused his attention on the historical part of Jorré's work, which appeared in print 15 years after the death of the author (Jorré 1971). Georges Jorré had been taught by Raoul Blanchard at the lycée in Douai in northern France, and then at the University of Grenoble. Jorré later taught at the Douai *lycée*; in 1929, he transferred to the *lycée* in Toulouse, and was an adjunct lecturer at the University from 1934. Using Russian language sources but not travelling to the Soviet Union, he completed a textbook, *L'URSS, la terre et les hommes*, in 1939; publication was refused by the censors during the Pétain-Laval regime. The book eventually appeared in 1946, and was translated into English by E. Laborde as *The Soviet Union: The Land and Its People* (Longman 1950). At the time of his death in 1956, Jorré had completed a research manuscript running to 1,251 closely-typed pages that would have formed his doctoral thesis. Faucher edited the first 11 of the 21 chapters, and these were published in 1971 as *Le Terrefort toulousain et lauraguais*. Final corrections were made by François Taillefer, following Faucher's death.

In this final stage of his life, Daniel Faucher moved to Nice to be close to his only daughter and her husband François Gay (1915–2000), who was professor of geography at the local university. However, he retained his house and his book collection in Toulouse, and had another property, Vaucourte, at Loriol-sur-Drôme, where he died on 17 July 1970 at the age of 88.

Geographical Ideas

Like other French contemporaries, Daniel Faucher was a 'complete geographer' with expertise in physical as well as human geography, and having the archival skills of an historian as well as competence in countless areas of human geography. His earliest academic interests had been in economic history but one of his earliest encounters with Raoul Blanchard involved going into the field to explore the terraces of the Rhône (Blanchard 1963, 142–3). Through Blanchard, Faucher received the holistic vision of geography conceived by Vidal and, as the only professor of geography at Toulouse, he was required to teach all aspects of the discipline for a number of years. His chosen specialism was *la géographie agraire* (rural geography) which enabled him to investigate the complex relationship between the natural and the cultural worlds, as mediated by different social groups through time as they employed varying cultivation practices, experimented with new crops and introduced new ploughs and machinery. Despite his rural bias, which he attributed to working in the fields alongside his grandmother, he also published on urbanization, glaciation, terrace formation, climate and many other themes. His favoured areas for research encompassed the Rhône valley and south-western France. He also taught and travelled in Iberia, and visited the Netherlands, Poland and the United States for international conferences. Towards the end of his life, he developed an interest in agricultural mechanization in the USSR, a topic that Russian-speaking Jorré had investigated for his textbook, *L'URSS* (Jorré 1946, 1950; Péchoux 1998). Not surprisingly, Faucher drew largely on French-language materials but he had a reading knowledge of English, Spanish and several other European languages. His fascination with North American modernity was shared by his young colleague, Taillefer, who is often remembered as 'an austere geomorphologist, but was remarkably open to other dimensions' of geography (Claval 1996, 22; J. Pilleboue, pers. comm. 6 December 2010). As François Gay would recall, 'whilst flying back from the USA in 1952, Faucher was struck by the extraordinary diversity of the landscape mosaic of provincial France, which was so different from the mechanized countryside and the monoculture on the plains of North America' (Gay 1982, 163).

Faucher's major thesis on the middle Rhône shared many of the characteristics of other French regional monographs, written by the first and second generations of Vidalians before 1914 and between the two world wars (Clout 2009). It began with an emotional dedication to Blanchard: 'He has been a teacher and a friend to me. He encouraged me so warmly that the disruption that the war brought to my studies was not fatal. My work owes more to him than he thinks; and I am more indebted to him than I know how to say' (Faucher 1927, xi). Running to 670 pages, it listed a formidable array of archival sources consulted in Paris, Valence and 4 other southern towns, made reference to over 400 publications, and was illustrated with 50 maps and diagrams, as well as a handful of photographs and field sketches. In addition, Faucher made use of case-study questionnaires and thanked 'the people who have responded, and among them a large number of village school teachers and pupil teachers' with whom contact was made while he taught at Valence (Faucher 1927, x). He recognized that 'the fragment of the Rhône valley' that he chose to study was both a territory of convergence and contact, and an area of north-south gradation and transition. Describing this section of the Rhône valley, he declared: 'We are placed where extremely varied geographical phenomena converge – or, more exactly, run together – as the waters of the Alps, of the Alpine foothills, of the Vivarais and of the Cévennes mix with those of the great watercourse of the Saône, which extends toward the Vosges, and is lost in the Mediterranean Sea to the south. We have thought that it would be useful to describe the plains and basins of the middle Rhône as a region of contact' (Faucher 1927, vii).

Faucher's methodology was characteristic of its time, seeking to trace 'a harmonious combination of all the factors present', which shaped the landscapes and economic activities of the valley (Faucher 1927, ix). 'We have wanted, as much as possible, to show how human adaptation to the environment has evolved through time. This is the only method which allows, not the discovery of geographical laws – since there are no real laws in human geography – but the geographical explanation of existing facts' (Faucher 1927, ix). His thesis duly comprised three sections, dealing with climate, relief and watercourses in three chapters, with historic aspects of human activity in seven chapters, and current features of human life in the remaining four chapters. It was distinctive in several respects. As befitted an area of transition, Faucher discussed climatic factors before analysing terrain; he also paid particular attention to past conditions, with the final section also covering agricultural activities during the first half of the nineteenth century. Discussion of population and settlement was placed right at the end of the thesis, following a lengthy analysis of changing economic conditions. Urban life, exemplified by Valence and Romans, was dismissed in only ten pages in contrast with his sustained attention to agriculture and the countryside. Introduction of innovations in farming practice was highlighted, such as mulberry bushes to support silkworms and new fodders to feed increasing numbers of livestock. Similarly, the importance of new communications was stressed, notably the coming of the railway that opened the middle Rhône to growing fruit and vegetables on a commercial scale.

Presenting the work of his disciple to the academic community, Blanchard praised his finesse, vigour and artistry. In writing 'this fine volume . . . I dare to say that Monsieur Faucher has succeeded completely' (Blanchard 1927, 466). Nonetheless, Blanchard stressed that more emphasis should have been given to variations in population density, and the situation in the Rhône valley should have compared with conditions in surrounding areas. He declared that the illustration of this volume 'is mediocre, maps and diagrams are reproduced poorly, and are scarcely legible . . . Here and there, interesting details would have merited fuller explanation. . . . Yet, the book might have been abbreviated' and discussion of terrain and of historic activities could have been consolidated (Blanchard 1927, 471). Despite these shortcomings, the author 'writes very well [and] there is not a single chapter that does not renew our knowledge of particular problems . . . This is a work of the first rank, which does honour to the French school of regional geography' (Blanchard 1927, 471).

André Allix, also a pupil of Blanchard's, praised the skill of Faucher in 'conveying the feeling of regional cohesion in the midst of so many diverse influences. Richly documented and displaying a very firm talent, [the author] gives a faithful picture of life' in the middle Rhône (Allix 1929, 206). Maximilen Sorre (1880–1962) (*Geographers* Vol. 27) noted that 'the South has been treated like a poor relation by French geographers, but now Monsieur Faucher has opened its door to us' (Sorre 1927, 232). His appraisal of the book was overwhelmingly positive, but he found the 'reasoning rather strained and laborious' in the discussion of rural life (Sorre 1927, 234). Lucien Gallois (1857–1941) (*Geographers* Vol. 24), veteran geographer from the Sorbonne, offered a largely factual account of this thesis originating from the rival 'school' of geography in Grenoble, but Marc Bloch (1886–1944) found it 'excellent', recommended it to fellow historians, and made interesting observations on the relationship between academic history and geography between the wars (Gallois 1928; Bloch 1929, 607). 'Many problems of human life in the past, which historians in other countries consider to be their exclusive hunting ground (*chasse gardée*) have been studied mainly in France by scholars trained as geographers. . . . I am absolutely unable to draw any substantial distinction between these disciplines whose common ambition, when one comes to the heart of the matter, is to build together, little by little, a science of mankind in society . . . Every economic historian should read this book, which is one of the richest and most solidly documented ever to have been written about regional change' (Bloch 1929, 607–8).

Faucher adopted a similar holistic and chronological approach in the final version of his *Géographie agraire*, with ten chapters tracing agricultural change, from gathering and slash and burn, through increasingly complex rotations, intensive cultivation and plant innovations, to orchards, vineyards, and industrial crops such as cotton, oil seeds and sugar beet (Faucher 1949). He explained how 'these cropping activities depend on the geographical environment but are also linked to a multitude of phenomena, which together, and in a particular place and at a particular time, represent a rural civilization' (Faucher 1949, 8). This book earned Faucher the Olivier de Serres Prize awarded by the Ministère de l'Agriculture, and in 1951 his students held a reception to commemorate this honour (P.-Y. Péchoux, pers. comm. 24 October 2010). It was, however, in *La vie rurale vue par un géographe* that Faucher fully articulated his methodological position and exemplified his research. He insisted that, while being subject to broad social and political changes, 'agriculture and rural life have their own rhythms, their own history', and that it was necessary to study 'the contact of mankind with natural elements that are used to obtain desired ends . . . Agriculture is a bio-technical complex . . . embracing mankind as well as the plants and animals that are brought into human service' (Faucher 1962, i). Thus, his use of the English term 'environment' covered both human and physical factors (Faucher 1962, iii). He placed rather less emphasis on field patterns and settlement patters than did Albert Demangeon (1872–1940) (*Geographers* Vol. 11), Marc Bloch, or Roger Dion (1896–1981) (*Geographers* Vol. 18), stressing that 'the geographer cannot just observe landscapes without going back to the psychological factors that influence human reactions to work. . . . Too few geographers, to my taste, interrogate the human sciences [of psychology, ethnology and sociology] in order to illuminate spatial distributions and to analyse relationships between groups' (Faucher 1962, iii). He placed less emphasis on practising 'a quantitative geography but, rather, [strove] to direct concern to a qualitative knowledge of the countryside' since, behind the production figures, 'it is the perceptions, beliefs and behaviour of the producer that must be reached' (Faucher 1962, iv).

The three dozen essays in *La vie rurale vue par un géographe* made accessible some of Faucher's unpublished conference papers and various contributions to minor, as well as major, periodicals. They showed that his chosen methodology was not very different from the integrated approach (*combinaison*) articulated in the 1940s by Cholley (*Geographers* this volume), his contemporary at the Sorbonne. Faucher displayed a real fascination for the garden plot, 'which is the most archaic element in agriculture, but is also the most progressive element', where countless innovative crops had been nurtured before being transferred to field production (Faucher 1962, 133). The garden, he believed, 'gives expression to the hidden personality of the farmer; his taste for work, his love of the land, that is worked, transformed and enriched, and laden with humanity' (Faucher 1962, 133). Faucher's research on rural life and landscapes in southern France provided an important complement to the findings of Bloch, Demangeon and Dion, who were more familiar with northern France. After evoking 'the broad blue sky' of le Midi,

Faucher insisted that it was 'in the clear pattern of the rocks, in the precise line of hills and basins, in the distribution of fields conquered by hard labour, in the arrangement of trees providing shade [that the southerner] finds himself and recognizes the fruits of his own effort, the link between himself and the forces of the land, and the legacy of his people from the past' (Faucher 1962, 218). In an essay drafted in 1944, he insisted that rural life in the South-West was 'closely adapted to physical factors and is not entirely in time with the present. It is the victim of a sort of slowness that gives it originality, and a strange taste for the land (*le terroir*), however that condemns it to a sort of languor. It will only be cured by a profound revision of its human status. It has experienced some initiatives, but it needs new ones. It retains the flavour of the farm shaped to the scale of family labour . . . It is still turned to the past; it must become aware of the necessities of the present. A rural civilization can only survive by changing; the land of south-western France, which has borne harvests for more than two millennia, needs to find a new soul (*une âme nouvelle*)' (Faucher 1962, 194).

In the immediate euphoria of post-war liberation, Faucher looked to the future and advocated the modernization of the countryside in many of his writings. A later essay of 1957 was particularly explicit on this point and anticipated fundamental legislation of 1960 and 1962 that would balance encouraging agricultural innovation in France with the need to keep progressive young farmers on the land. He declared:

> The problem is to train young farmers who will understand that this is not the time for narrow conservatism, and who will show what can be expected from . . . a renovated farm structure, and from an agriculture that is open to progress . . . It is up to the whole nation to prepare the way for this transformation of farming life. It has often been written that the machine liberates the worker. It would be more exact to say that mankind is required to put machines fully into their service. Today, the dilemma is posed thus: we must innovate or die. And to die would be a disaster. We must keep young people on the land and identify the courageous young folk who will innovate otherwise the whole social equilibrium will be broken rapidly. [Mass urbanward migration] will open the way toward a crisis in moral and intellectual values that are essential to French civilization. It is the close contact of people with the land that forges the qualities of common sense, finesse, calm courage, and hope . . . Our rural youth knows well that our future is in their hands. (Faucher 1962, 301)

As he advanced in years, Faucher looked increasingly towards the challenges of the future and the need to innovate and change, both in the French countryside and along the banks of the Rhône, but never lost sight of experience from the past and the many legacies of earlier times to the present (Wolff 1971, 10).

Influence and Spread of Ideas

After training future primary teachers at Valence for more than a decade, and co-authoring three school textbooks with Blanchard, Daniel Faucher pioneered the teaching of modern geography at the University of Toulouse, adopting a friendly, even familial approach to his students, much as his master had done at Grenoble (Blanchard and Faucher 1923, 1924, 1925). Like Blanchard, he established a 'geographical laboratory' and inaugurated a geographical periodical that he edited for over 30 years. He trained countless *lycée* teachers and encouraged some to undertake the equivalent of Masters research. A few of these disciples proceeded to doctoral work and obtained posts in higher education. These advanced scholars were important contributors to the *Revue Géographique des Pyrénées et du Sud-Ouest* whose articles transformed geographical knowledge of south-western France. Their affection for their professor was expressed through the two-volume jubilee volume, *France méridionale et pays ibériques*, and through their support for *La vie rurale vue par un géographe* (Anon. 1948–9; Faucher 1962). In many respects, Faucher's activity was not greatly different from that of the small number of professors of geography who taught in French provincial universities before and immediately after World War II. There were, however, some differences. Faucher developed his own conception of geography 'in defiance of Parisian geography' (J. Pilleboue, pers. comm. 6 December 2010). He master-minded *La France, géographie, tourisme*, which provided a comprehensive, up-to-date regional text that appealed not only to academics and *lycée* teachers, but also to many members of the book-purchasing public (Faucher 1951–2). Its chapters revealed that modern geography was not about tedious lists of place names and production statistics but was a vibrant science that brought together physical and cultural features to explain both the uniqueness of place and the appeal of certain places to tourists. His *Géographie agraire* provided a comprehensive student textbook that, once again, filled a gap in the existing literature; its line of reasoning was continued by *Le paysan et la machine* and *La vie rurale vue par un géographe.* His collaboration with Georges Duby and Isaac Chiva ensured that geographical research was welcomed in the pages of *Etudes Rurales*.

There is, however, a further dimension of the influence of Daniel Faucher, whose name was given in 1966 not only to the Institut de Géographie of the University of Toulouse, but also to the complex of student residences he promoted and managed (Anon. 1966). The destination 'Daniel Faucher' appears on buses serving these residences but it is unlikely that passengers are aware of its significance (P.-Y. Péchoux, pers. comm. 24 October 2010). Faucher's name was also attached to various schools in south-western France and in the Rhône valley. As well as being an academic, he was an engaged citizen who assumed countless responsibilities. While at Valence, he was active in anti-fascist organizations and in the League for Human Rights, invited Albert Thomas (former secretary of state for armament), and a group of German teachers to visit the town immediately after the conclusion of World War I, and was president of the local branch of the teachers's league (Taillefer 1970, 215). At Toulouse, he encouraged vacation courses for serving teachers and supported the provision at Burgos of Spanish language courses for Toulouse students (Wolff 1971, 7). He strove to ensure that facilities were provided to house foreign students and those fleeing the Spanish Civil War. During World War II, he condemned the geopolitical concept of 'lebensraum' behind the German invasion of territories in continental Europe and, together with philosophers Etienne Borne and André Dumez (aka Vladimir Jankélévitch), wrote a 20-page pamphlet for the national movement against racism entitled 'The racist lie' (Faucher 1940a; Faucher, Borne and Dumez 1944). He supported a Resistance network, and used his influence to protect Jewish students and to enable male students to avoid compulsory labour service in Germany. As soon as German invasion of the southern 'free zone' seemed likely, he transferred much of the library holding of his beloved institute to the municipal library in Toulouse for safe keeping (P.-Y. Péchoux, pers. comm. 24 October 2010). From 1942, he was involved with the Bureau du Régionalisme and developed further contacts with industrialists, foresters and hydrologists, as well as with farming associations in south-western France.

Later, as dean, he planned and managed student welfare services at Toulouse. He was active in the regional association for teachers of history and geography, and was president of the Fédération des Sociétés Savantes des Pyrénées-Languedoc-Gascogne that gave its support to local learned societies. Until his retirement in 1952, he was vice-president of the council of the University of Toulouse, and in May 1968 he reappeared for what proved to be the final meeting of the old Faculté des Lettres to plead for calm and reason. That was the time when responsibilities were transferred to a new management committee, comprising representatives of the student body and of young lecturers, as well as established professors (P.-Y. Péchoux, pers. comm. 24 October 2010). As his historian colleague Philippe Wolff declared, Faucher 'crossed the storm of May 1968 with benevolent attention, being full of common sense' (Wolff 1971, 8). For short periods, Faucher was a member of the national centre for scientific research and was chairman of the commission for agricultural geography of the International Geographical Union. In retirement, he directed the regional pedagogic institute at Toulouse, and encouraged the University to support a new periodical, the *Annales de la Faculté des Lettres de Toulouse* (P.-Y. Péchoux, pers. comm. 24 October 2010). He was founding chairman of the Comité d'études pour l'aménagement de la region toulousaine that advocated not only a national system of regions, of which 'Midi-Pyrénées' (surrounding Toulouse) would be an integral part, but also the allocation of regional development funds to counter-balance the concentration of employment, services and wealth in Paris and its immediate surroundings. The subsequent modernization, industrialization and economic buoyancy of Toulouse and its environs owed much to these efforts. For his advocacy of rural and regional development policies, he was appointed a corresponding member of the Académie des sciences morales et politiques of the Institut de France in 1954. Prior to retirement, his office door was always open to visitors, since he worked on research, writing and editing at home, in the early morning and late into the evening. As François Taillefer recalled: 'Before being the professor, or the dean, he was himself . . . He lived for other people, for his students and all the students of the University, but also for all those who needed help or advice' (Taillefer 1970, 217). Even in retirement, this 'good and generous' man made known his availability at his home address (Taillefer 1970, 218).

Six decades after his retirement and four decades after his death, the academic work of Daniel Faucher has been largely forgotten. Even in the late 1960s, it was rarely mentioned in the Institut de Géographie that by then bore his name. As lecturer Jean Pilleboue noted, 'History was indeed cruel' since he was still alive (J. Pilleboue, pers. comm. 12 October 2010). The new generation of rural researchers in the University of Toulouse:

> had read his main publications, but all thoughts were directed to charting the frenetic changes that were taking place in countryside. For the South-West, we followed the work of very prolific authors, such as Roger Brunet (b. 1931) and Bernard Kayser (1926–2001) . . . At that time, there was a feeling – surely erroneous – that Faucher's analyses did not help us to understand the dramatic changes taking place in the 1960s and 1970s . . . When Faucher was remembered, it was more as a very engaged citizen than as a rural geographer. That was not unfair; he had been in the Resistance before the Germans occupied southern France [in 1942], he had helped Spanish Republican refugees in Toulouse, and he continued his civic involvement after the war. Bernard Kayser [appointed professor at Toulouse in 1958], who had been in the Resistance, greatly admired the active citizenship of Daniel Faucher. (J. Pilleboue, pers. comm. 12 October 2010)

His name survives on the marble plaque at the entrance to the 'Institut de Géographie Daniel Faucher' on the campus of Le Mirail. During a prolonged strike in 1976, some rather talented students drew a large hammer on the surrounding brick wall and chalked a jagged line across the plaque. For Jean Pilleboue, 'this signified respect for Faucher – since the plaque was in no way damaged – but also the fact that the passage of time had, in some way, split his work apart' (J. Pilleboue, pers. comm. 12 October

2010). He was 'a talented representative of a form of geography whose main concern was to identify itself as a distinctive scientific discourse . . . After him, geography in France became disturbed and dissatisfied' with its previous manifestations (J. Pilleboue, pers. comm. 6 December 2010). The legacy of Daniel Faucher, albeit now overlooked if not forgotten, may be found in his work as a teacher, his role as a scholar and his many contributions as a socially engaged citizen in his city of adoption (P.-Y. Péchoux, pers. comm. 24 October 2010).

Conclusion

As for many bright young people of humble origin, a career in school teaching provided a means of social advancement for Daniel Faucher. Unlike most of those men and women who remained teaching in the villages and towns of France, however, he became a university professor and an influential dean of faculty. Completing his doctorate at the University of Grenoble, rather than at the Sorbonne in Paris, he contributed to the development of geographical expertise in the French provinces. Once installed at Toulouse, he encouraged geographical research in a rather overlooked part of the country, and gave his support to a small group of doctoral scholars. South-western France was not his exclusive preserve, since several researchers studying aspects of the Pyrenees and the great valleys extending from them submitted their theses at the Sorbonne. The influential study by Pierre Deffontaines (*Geographers* Vol. 30) on the middle Garonne valley was but one example among several (Deffontaines 1932). Faucher's vision of geography was both conventional and prophetic. He embraced the contributions of history, geology and botany to the discipline but he also stressed the importance of the human sciences to enhancing geographical understanding. It would be exactly this kind of recognition of innovations in sociology, political studies and other social sciences that would stimulate the activities of a new generation of scholars at the Institut de Géographie- Daniel Faucher following his death.

Acknowledgements

My sincere thanks to Pierre-Yves Péchoux and Jean Pilleboue in Toulouse, and to Hugh Prince in London for encouragement and information.

Bibliography and Sources

1. OBITUARIES AND OTHER ACCOUNTS RELATING TO DANIEL FAUCHER

Allix, A. (1929), 'Plaines et bassins du Rhône moyen entre Bas-Dauphiné et Provence', *Etudes Rhodaniennes* 5, 202–6.

Anon. (1948–9), *France méridionale et pays ibériques. Mélanges géographiques offertes à Daniel Faucher,* 2 vols. Toulouse: Editions de l'Ingénieur and Privat.

Anon. (1966), 'La vie et le rayonnement de la Faculté. Cérémonies en l'honneur du Doyen Daniel Faucher, 21 Octobre 1966', *Bulletin de la Faculté des Lettres et Sciences Humaines de l'Université de Toulouse* 3, 241–8.

Blanchard, R. (1927), 'Plaines et bassins du Rhône moyen entre Bas-Dauphiné et Provence', *Revue de Géographie Alpine* 15, 465–71.

Blanchard, R. (1963), *Je découvre l'université: Douai, Lille, Grenoble*. Paris: Fayard.

Bloch, M. (1929), 'Une monographie géographique: les pays du Rhône moyen', *Annales d'Histoire Economique et Sociale* 1, 606–11.

Brunet, R. (1965), *Les campagnes toulousaines*. Toulouse: Faculté des Lettres de l'Université de Toulouse.

Clout, H. (2009), *Patronage and the Production of Geographical Knowledge in France: The Testimony of the First Hundred Regional Monographs, 1905–1966*. London: Historical Geography Research Series, No. 41.

Deffontaines, P. (1932), *Les Hommes et leurs travaux dans les pays de la moyenne Garonne*. Lille: SILIC.

Gallois, L. (1928), 'Le couloir du Rhône entre Bas-Dauphiné et Provence', *Annales de Géographie* 37, 177–180.

Gay, F. (1982), 'Daniel Faucher, géographe du monde rural', *Revue Géographique des Pyrénées et du Sud-Ouest* 53, 161–6.

George, P. (1935), *La Région du Bas-Rhône*. Paris: Baillière.

George, P. (1969), 'L'homme et le Rhône', *Annales de Géographie* 78, 718–721.

Gibert, A. (1950a), 'Les mélanges Daniel Faucher', *Revue de Géographie de Lyon* 25, 77–9.

Gibert, A. (1950b), 'Sur la géographie agraire générale', *Revue de Géographie de Lyon* 25, 315–17.

Grivot, F. (1971), 'Bibliographie des travaux de Daniel Faucher', *Annales de Géographie* 80, 388–96.

Jorré, G. (1971), *Le Terrefort toulousain et lauraguais*. Toulouse: Privat.

Juillard, E. (1971), 'Daniel Faucher, 1882–1970: l'œuvre', *Etudes Rurales* 41, 11–14.

Papy, L. (1971), 'Daniel Faucher, 1882–1970', *Annales de Géographie* 80, 385–8.

Péchoux, P.-Y. (1998), 'De trois géographes liés à Toulouse et de leurs géographies de la Russie', *Slavica Occitania* 7, 183–202.

Péchoux, P.-Y. (2003), 'Jean Sermet, 1907–2003', *Bulletin de la Société de Géographie de Toulouse* 283, 3–5.

Péchoux, P.-Y. (2006), 'François Taillefer, 1917–2006', *Bulletin de la Société de Géographie de Toulouse* 286, 4–7.

Robert, J.-B. and Allix, A. (1954), 'La France, géographie, tourisme', *Revue de Géographie de Lyon* 29, 154–6.

Robic, M.-C. (1984), 'Daniel Faucher, 1882–1970', in P. Pinchemel, M.-C. Robic and J.-L. Tissier (eds), *Deux siècles de géographie française: choix de textes*. Paris: Comité des Travaux Historiques et Scientifiques, 206–7.

Sorre, M. (1927), 'Plaines et bassins du Rhône moyen entre Bas-Dauphiné et Provence', *Bulletin de la Société de Géographie de Lille* 69, 231–5.

Taillefer, F. (1970), 'Daniel Faucher', *Revue Géographique des Pyrénées et du Sud-Ouest* 41, 213–18.

Taillefer, F. (1975), 'Daniel Faucher, 1882–1970', in P. George (ed.), *Les géographes français*. Paris: Comité des Travaux Historiques et Scientifiques, 173–83.

Wolff, P. (1971), 'Daniel Faucher, 1882–1970: l'homme', *Etudes Rurales* 41, 7–10.

2. *SELECTED WORKS BY DANIEL FAUCHER*

1914 'La plaine de Valence', *Annales de Géographie* 23, 127–50.

1918a 'Les forets des montagnes de la Drôme à la fin du XIXe siècle et de nos jours', *Receuil des Travaux de l'Institut de Géographie Alpine* 6, 237–53.

1918b 'En Macédonie: le Tikvès', *Annales de Géographie* 27, 226–30.

1919 'En Macédonie: le Gauda ', *Annales de Géographie* 28, 471–5.

1920 'Le pays de Quint, Diois septentrional', *Revue de Géographie Alpine* 8, 259–326.

1923 (with R. Blanchard) *Cours de géographie. L'Europe, Les grands pays du monde.* Paris: Gédalge.

1924 (with R. Blanchard) *Cours de géographie. La France et ses colonies.* Paris: Gédalge.

1925 (with R. Blanchard) *Cours de géographie. Cours élémentaire.* Paris: Gédalge.

1927 *Plaines et bassins du Rhône moyen entre Bas-Dauphiné et Provence.* Paris: Armand Colin.

1931 'Le maïs en France', *Annales de Géographie* 40, 113–21.

1932 'Le glacier de l'Ariège dans la basse vallée montagnarde', in *Mélanges géographiques offerts à Raoul Blanchard.* Grenoble, 183–99.

1933 'Les lacs des Pyrénées françaises et la morphologie glaciaire pyrénéenne', *Revue Géographique des Pyrénées et du Sud-Ouest* 4, 216–29.

1934 'Polyculture ancienne et assolement biennal dans la France méridionale', *Revue Géographique des Pyrénées et du Sud-Ouest* 5, 241–55.

1935 *Géographie agraire. Types de culture.* Lisbon: Documentation économique française (Second edition, 1940).

1936 'Les conditions naturelles de la vie agricole en Grèce', *Bulletin de la Société Toulousaine d'Etudes classiques* 9, 1–45.

1936 'Surfaces d'aplanissement dans les Pyrénées ariégeoises', in *XIVe Congrès International de Géographie, Varsovie, 1934: Comptes-Rendus*, vol. 2. Warsaw: Union Géographique Internationale, 472–6.

1936 'Toulouse, capitale régionale', in *XIVe Congrès International de Géographie, Varsovie, 1934: Comptes-Rendus*, vol. 3. Warsaw: Union Géographique Internationale, 190–7.

1937 'Glaciaire pyrénéen et vallées sous-pyrénéennes', in *Mélanges géographiques et orientalisme offerts à E.F. Gautier.* Tours: Arrault, 213–25.

1937 'Le glacier de l'Ariège dans la basse vallée montagnarde', *Revue Géographique des Pyrénées et du Sud-Ouest* 8, 335–49.

1938 'Surfaces d'aplanissement dans les Pyrénées ariégeoises', in *Xve Congrès International de Géographie, Amsterdam, 1938: Comptes-Rendus, Commission pour la cartographie des surfaces d'aplanissement tertiaires*. Leiden: Brill, 34–8.

1940a 'Espace vital et géographie', *La Grande Revue* 1940(2), 3–20.

1940b 'L'agriculture des Pyrénées françaises. Caractères généraux', *Revue Géographique des Pyrénées et du Sud-Ouest* 11, 39–54.

1940c 'L'unité du Portugal', *Revue Géographique des Pyrénées et du Sud-Ouest* 11, 159–66.

1940 (with C. Higounet, J. Sermet, R. Plandé) 'Les villes des l'Aquitaine orientale', *Revue Géographique des Pyrénées et du Sud-Ouest* 11, 81–110.

1942 'Les pré-Pyrénées ariégeoises et garonnaises. Essai d'étude morphogénétique d'une lisière de la montagne', *Revue Géographique des Pyrénées et du Sud-Ouest* 13, 358–65.

1942 'Villefranche-de-Rouergue et la Haute-Garonne dans la nouvelle organisation provinciale', *Revue Géographique des Pyrénées et du Sud-Ouest* 13, 263–74.

1943 'Les Pyrénées: aperçu géographique', *Revue Géographique des Pyrénées et du Sud-Ouest* 14, 204–36.

1944 (with E. Borne and A. Dumez, aka V. Jankélévitch) *Le mensonge raciste; ses origines, sa nature, ses méfaits*. Toulouse: Mouvement national contre le racisme (reprinted in J. Estèbe, *Les Juifs à Toulouse et en Midi toulousain au temps de Vichy*. Toulouse: Presses Universitaires du Mirail, 1996).

1945 'Evolution des types de maisons rurales', *Annales de Géographie* 53–4, 241–53.

1949 *Géographie agraire. Types de cultures*. Paris: Médicis (translated as *Geografia agraria. Tipos de cultivos*. Barcelona: Omega, 1953).

1949 'Le Languedoc et les civilisations rurales méditerranéennes', *Annales de l'Institut d'Etudes occitanes* 2, 3–13.

1951–2 (as editor) *La France: géographie, tourisme,* 2 vols. Paris: Hachette.

1953 'Evolution récente du genre de vie dans les campagnes françaises', in Centre d'Etudes Sociologiques, *Villes et Campagnes, civilisation urbaine et civilisation rurale en France*. Paris: Armand Colin, 363–400.

1954 'De quelques incidences en France de la 'révolution agricole' du XVIIIe-XIXe siècle', in *Mélanges géographiques offerts au Doyen E. Bénévent*. Gap: Ophrys, 111–24.

1954 *Le Paysan et la machine*. Paris: Editions du Minuit.

1954 (with M. Chevalier) 'Immigrations et progrès agraire: des colons tunisiens dans les campagnes toulousaines', *Annales, Economies, Sociétés, Civilisations* 9, 347–55.

1956 'Georges Jorré', *Revue Géographique des Pyrénées et du Sud-Ouest* 27, 195–6.

1956 *La Machine dans l'agriculture française*. Paris: Documentation Française.

1956 'La révolution agricole du XVIIIe-XIXe siècle', *Bulletin de la Société d'Histoire moderne* 55, 2–11.

1957 'L'activité géographique française: les revues', in G. Chabot, R. Clozier and J. Beaujeu-Garnier (eds), *La Géographie française au milieu du XXe siècle.* Paris: Baillière, 299–304.

1959 'Les jardins familiaux et la technique agricole', *Annales, Economies, Sociétés, Civilisations* 14, 297–307.

1959 'Réflexions sur le destin de Toulouse', *Revue Géographique des Pyrénées et du Sud-Ouest* 30, 101–15.

1961 'L'assolement triennal en France', *Etudes Rurales* 1, 7–17.

1962 *La Vie rurale vue par un géographe.* Toulouse: Institut de Géographie.

1968 *L'Homme et le Rhône.* Paris: Gallimard.

The publications of Daniel Faucher, which exceed 250, are listed in *La vie rurale vue par un géographe*, and in the *Annales de Géographie* (1971), 388–96.

Chronology

1882	Born on 3 February at Romans (Drôme)
1898–1903	Trained as pupil-teacher at Valence
1903	Appointed village schoolmaster at Loriol-sur-Drôme
1907	Married Jeanne Caillet, a fellow teacher
1912–26	Taught at teacher training college in Valence, while taking a degree and then undertaking doctoral research at the Institut de Géographie Alpine, University of Grenoble
1914–18	Served in French army, first in the Vosges, then in Macedonia
1926	Appointed to the University of Toulouse, to succeed Paul Dognon
1927	Defended doctoral thesis, *Plaines et basins du Rhône moyen entre Bas-Dauphiné et Provence*
1930	Founded the *Revue Géographique des Pyrénées et du Sud-Ouest*; editor until 1964
1935	Lecture series on 'La géographie agraire' at Lisbon
1939	Inauguration of Institut de Géographie, rue de l'Université
1942	Appointment of François Taillefer as Faucher's *assistant*
1944–52	Elected Dean of the Faculté des Lettres; twice re-elected
1948	Celebration of Faucher's work at Toulouse; publication of *La France méridionale et pays ibériques*
1949	Publication of *La France agraire*

1950	Death in late September of Madame Jeanne Faucher
1952	Retired from University of Toulouse; succeeded as professor of geography by François Taillefer
1954	Publication of *Le Paysan et la machine*
1961	Appointed joint editor of *Etudes Rurales*
1962	Publication of *La Vie rurale vue par un géographe*
1968	Publication of *L'Homme et le Rhône*
1970	Dies on 17 July at Loriol-sur-Drôme

Kenneth Cumberland

1913–2011

Eric Pawson

Kenneth Cumberland was a Yorkshireman who became the first geographer 'by qualification' to teach the subject at university level in New Zealand. The epithet was bestowed by George Jobberns (*Geographers* Vol. 5) when Cumberland joined him, aged 24, in the new Department of Geography at Canterbury University College in Christchurch in May 1938. To Cumberland the dramatic and active landscape of the South Island soon became 'a geographical laboratory . . . an almost virgin and truly pioneer sphere for the geographer' (Cumberland 1940, 19–20). He quickly established a position for himself as a prolific author on the theme that came to define his research career, that of the wise use of resources. In 1946, he moved to Auckland University College to set up its department, and was appointed to the newly created chair in 1949. He produced a succession of regional texts for use in schools and by undergraduates, and had a strong sense of the civic purpose and wider application of geography. He gave public radio broadcasts for 30 years, was active in local body politics, and served 2 terms as a Vice President of the International Geographical Union. In his retirement he both scripted and fronted a ten-part television documentary series, *Landmarks*, first shown in 1981, for which he is still widely remembered by older New Zealanders.

Education, Life and Work

Cumberland was born on 1 October 1913 in Bradford, Yorkshire, the elder son of Ernest Cumberland and Lucy (née Hodgson). Ernest was appointed as cashier with Charles Sykes and Company, manufacturers of knitting yarns, during World War I, remaining with the firm for 38 years and rising to mill manager and then company secretary. Ernest's father and mother had moved to Bradford in 1871 from rural Leicestershire. Lucy's family were also rural people, from Kettlewell-in-Craven in Wharfedale. It was Kettlewell rather than Bradford that was the emotional anchor for Kenneth throughout his childhood and student days. Its patterns of land utilization were the basis of the thesis work for his MA, forming the first step in his research career. On return visits from New Zealand, even many years later, Kettlewell drew him back.

Cumberland was educated at Grange Elementary and Grange High Schools in Bradford. He became school captain in his last year in the sixth form, as well as captain of both its football and cricket elevens. On the soccer field he played at centre half; on the cricket pitch he was a left arm leg break bowler. This was during Yorkshire cricket's heyday: he played against Len Hutton when both were at school; Herbert Sutcliffe and Percy Holmes were then opening for the county. Sport was for long Cumberland's main interest, and he had been in the soccer, cricket and swimming teams for his house since Form 1. In 1929, he was invited to join the Bradford Dolphins Swimming Club, where he also played water polo. Not until his first year in the sixth form did his academic interests begin to rival those in sport. He had taken geography, history, German and art, along with English and mathematics, for matriculation. For higher school certificate, he selected geography, history and German. He was the first boy from Grange to take geography at this level. He attributed his interest in the subject to the clarity of the landscapes in the vicinity of Kettlewell, as well as to the experience of travelling around the north of England with the soccer, cricket, swimming and water polo teams. His first ventures abroad were on school exchanges to Germany in 1930 and 1931.

Between the two world wars, a number of university colleges were opened in England to teach candidates for University of London external degrees. One of these was in Nottingham at which Cumberland took up an Open Exhibition in October 1932. Here he continued with sport, as a member in the university water polo team, representing the college in the Universities' Athletic Union (UAU) championships and on occasion playing for the county. In July 1933 he was selected by the UAU to represent the British Universities in the freestyle sprints and water polo at the International University Games in Turin. Jack Lovelock, the New Zealand miler, then at Oxford, was also in the team. These activities took their place alongside study for his degree. The first year was an intermediate of four subjects, for which Cumberland again chose geography with history and German, to which was added compulsory Latin. The two years of the degree course proper consisted of papers in physical and human geography, five in regional geography, and two electives, geomorphology and economic geography. Much of his teaching was with K. C. Edwards. He 'read quite a lot of geography in German' (Cumberland 2011, 72) and paid his last pre-war visit to Germany in the mid-1930s, when he acted as interpreter for a Nottingham field party to Thüringian Forest, a district later to be isolated behind the Iron Curtain. Years later he recalled that 'The most enjoyable, and indeed the most profitable and rewarding part of the geography course at Nottingham was the fieldwork we were required to do' (Cumberland 2011, 82).

Having taken his degree with First Class Honours, Cumberland was awarded a Cropper Postgraduate Fellowship to support research towards a two-year Masters of Arts programme. This he began in October 1935, under the supervision of Ken Edwards. His thesis title was 'The Agricultural Geography of Craven'. The Land Utilization Survey of Britain headed by Dudley Stamp (*Geographers* Vol. 12) and E. C. Willatts had not yet covered this remote part of Yorkshire and the thesis was designed to collect the requisite field data. Willatts provided Cumberland with cadastral map sheets in exchange for this. Jim Fox, a close Nottingham friend, gave a great deal of help in the field. In addition to the Land Utilization Survey, Sir George Stapledon's *The Land: Now and Tomorrow*, also published in 1935, was a source of inspiration. Stapledon was a well-known grassland scientist and advocate for 'wise use' of land and the future of the countryside.

In 1936, the British Association for the Advancement of Science met in Norwich. Edwards encouraged Cumberland to attend, where he met Professor C. B. Fawcett (*Geographers* Vol. 6) of University College, London. Fawcett had been his external examiner the year before; within a week a letter arrived offering Cumberland appointment as a temporary junior lecturer at University College in what was then one of the best Geography departments in the country. With a year of his research degree course yet to run, but mindful of the scarcity of academic positions in the aftermath of the Depression, he accepted the position, which involved teaching first year practical classes and a third year course on Asia in London. In addition to Fawcett, his colleagues were R. E. Dickinson (*Geographers* Vol. 8), Arthur Smailes and R. Ogilvie Buchanan, an economic geographer from New Zealand, who was well published on its primary industries. Cumberland was able to complete his Yorkshire fieldwork, and in 1937 presented a paper on the research to the Geography section of the British Association meeting. This time it was held in Nottingham. On the basis of a positive reception from geographers such as Stamp and Fawcett, Cumberland submitted a revised paper to the *Scottish Geographical Magazine*. It was accepted, appearing as his first publication in March 1938.

Earlier in 1938, Buchanan had received a letter from George Jobberns, asking if he could help to recruit a lecturer to assist in the new Department of Geography at Canterbury University College. A similar letter that Jobberns sent to Carl Sauer (*Geographers* Vol. 2) at Berkeley was delayed in the mail. Sauer responded, recommending Robert Bowman, son of Isaiah, the famous American geographer who was by then President of Johns Hopkins University (on Isaiah Bowman, see *Geographers* Vol. 1). But the reply was too late (although Bowman was soon to visit Canterbury anyway). Buchanan had showed his letter to Cumberland straight away, and the latter made up his mind 'within a matter of hours, that if Canterbury College was prepared to make it possible, I was going to go to New Zealand' (Cumberland 2011, 95). He boarded a ship at Tilbury a few weeks later, on 26 March 1938.

The sea voyage to New Zealand at the time took in Gibraltar and other Mediterranean ports, Port Said, the Suez Canal and Aden, then crossed the Indian Ocean to Colombo, before calling at Fremantle, Adelaide, Melbourne and Sydney. In Melbourne, a geologist from the university showed Cumberland around the urban area. In Sydney, he dined with Professor James Macdonald Holmes (*Geographers* Vol. 7), the head of the only Geography department in Australia. He sailed in a second ship across the Tasman to Wellington, and in a third to Lyttelton, the port town for Christchurch. The journey finished with a short train journey through the tunnel under the Port Hills to Christchurch railway station. There, under the clock, Dr George Jobberns was waiting to meet him.

Jobberns had established the Geography department at Canterbury University College in 1937. He had been a schoolteacher and teachers' college lecturer. He had an MA in Geology (and was awarded his DSc in 1936) but described himself as 'a geographer by declaration' (Roche 2010). It was the prospect of the second intake to the new department, and of the first intake moving through to a second year, that had prompted Jobberns to seek the assistance of a new lecturer. To Cumberland, Jobberns was 'a kindly, but wily, soul'. He impressed the younger man in seeming 'to know personally almost everybody in the country' (Cumberland 2011, 109). New Zealand was a small place, with a population of only 1.6 million people, barely a third that of Yorkshire. And Jobberns had influential networks as both a freemason and a member of the Canterbury Club (Cumberland 1975; Soons 2010). He saw Cumberland as '[o]ne of the ablest of the crop of young English graduates' (in Roche 2010, 92), but also as 'a shy, diffident, and in some ways difficult English boy in his mid-twenties' (Roche 2010, 103). Cumberland however quickly found his feet. Taking his cue from Jobberns, he showed no diffidence in embarking on his new career.

A good description of the intellectual atmosphere of New Zealand in the late 1930s has been left by the philosopher Karl Popper, who arrived at Canterbury University College from Vienna a year before Cumberland. Popper spent nine years in Christchurch and found the people 'decent, friendly, and well disposed'. But '[i]n those days New Zealand had no contact with the world except through England, five weeks away. There was no air connection and one could not expect an answer to a letter in less than three months'. This made for 'a wonderfully quiet and pleasant atmosphere for work . . . I won a number of friends who were interested in my work and who greatly encouraged me' (Popper 1976, 112). He in turn was greatly to encourage Cumberland. From his second year, Cumberland was allocated a room next to Popper's above the college workshop. With Popper he discussed Richard Hartshorne's *The Nature of Geography* when it was published in 1939. Hartshorne's commitment to tracing the origins of the modern discipline in the German authors – Kant, Hettner, Humboldt and Ritter – appealed to Cumberland, as did its contribution towards stimulating 'the emerging coherence of world geography as an intellectual discipline, at least in the English speaking realms' (Stoddart 1989, 163).

Cumberland's initial teaching duties were Stage I practicals, including map work, and a Stage II course on 'the regional geography of Europe, including the British Isles'. Jobberns was happy to widen his Stage I course to include human geography, and then teach higher-level physical geography. At the time New Zealand school syllabi in geography were restricted to 'elementary physiography', 'mathematical geography', including latitude and longitude, and 'political geography – the political subdivision of the continents, boundaries and frontiers' (Cumberland 2007, 64). They were infused with environmental determinism (Cumberland 1956b; Roche 2011). In this context what the two were doing was very different, and Cumberland was later to refer specifically to the Jobberns' 'enthusiasm for a "humanised" geography' (1975, 2). He derived this in part from the Berkeley school (Roche 2011). Not only was Jobberns prepared to back the teaching of human geography, he also believed that geographers should help shape national debates about natural resource management and land use planning (Jobberns 1945; Roche 2011). Jobberns encouraged the young Cumberland in both these ways.

In 1939, Jobberns left for a nearly year-long visit to the United States, financed by a Carnegie travel grant. His teaching was covered by Robert Bowman, whose services had been secured as a visiting lecturer. Bowman's interest was in soil erosion. On this trip, Jobberns visited a number of geographers, including Carl Sauer, Bowman's supervisor at Berkeley. On Jobberns' return to New Zealand in mid-1940, he wrote to Sauer that 'I arrived back without incident to find Bowman and Cumberland well under way with the year's work and evidently very much enjoying it' (in Macaulay 1987, 40). In 1941, Bowman was succeeded by another of Sauer's students, Andrew Clark, an historical geographer (*Geographers* Vol. 14). Cumberland got on particularly well with Clark, later writing that '[h]e and I had long and valuable discussions on Hartshorne's exposition of the historical and logical development of the subject . . .' (Cumberland 2011, 135). Jobberns considered that 'these contacts had a mellowing and maturing effect on Cumberland and contributed a lot to his academic development' (in Roche 2010, 103–4).

Two aspects of Cumberland's career that now began to loom large were his contributions to establishing a public role for geography at university level and within schools, and his burgeoning research interests (Pawson 2011). He worked with Jobberns to secure the admission of geography at Canterbury to the Faculty of Science and its approval as a course of study for the BSc and MSc degrees. Opposition came from physicists and chemists who feared competition for limited equipment funds from a new subject. 'It was this antagonism that later required my election to the professorial board as a lecturers' representative. George organized that' (Cumberland 2007, 64). The Entrance Board of the University of New Zealand was responsible for school curricula. The key to securing curriculum change was the academic members of this Board, who were representatives of the professorial boards of the four university colleges (at Auckland,

Wellington, and Otago in Dunedin, as well as Canterbury). A later assessment argued that although Cumberland was 'probably the leading reformer of the geography curriculum', the initiative was more broadly based (McCulloch 1992, 170).

In his memoir, Cumberland was to claim that they were successful in making the content of the old school syllabus 'incidental to the teaching of the regional geography of those countries in the world that had some relevance to New Zealand' (2011, 114). Jobberns himself had been a producer of school texts (Roche 2011), but lacking training as a human geographer, gave way to Cumberland's vigorous entry into this field. The latter first produced an atlas for pupils in schools in New Zealand and Australia that went through 13 editions between 1942 and 1957. A third of its space was devoted to the two countries, and Europe, especially the British Isles, was also 'given special attention' (Cumberland 1942; 1956 edition, 2). A year after the first edition of the atlas, his *Map Reading in Geography* appeared, at a time when New Zealand lacked publicly accessible topographic maps. Its intent was to provide the materials to improve student map literacy in the absence, but in anticipation, of the topographic series that began to appear after World War II (Cumberland 1943a).

Cumberland was called up for war service but was declined on grounds of poor eyesight. Instead he spent the war years attending to his increasingly busy research career and family life. He had been introduced to much of Canterbury on trips with Jobberns and on class fieldwork, often in fairly primitive conditions. Many roads were unsealed and groups of students walked long distances and stayed in tents and huts. Cumberland's framework of interpretation in the field and classroom was Sauer's 'The morphology of landscape', which had appeared in 1925, and in which it was argued that 'area or landscape is the field of geography' (1963, 316). His own first paper on New Zealand discussed the regional landscapes of Canterbury, which were dramatically different to anything he knew in Britain (Cumberland 1940). He was becoming particularly familiar with the dry high country, an extensive district of mountain basins and valleys west of Christchurch, in the lee of the Southern Alps. It was to provide much of his insight into land use problems, such as overstocking, weed infestation and accelerated erosion. The 1940 paper anticipated a number he wrote in the *Geographical Review* that decade. The second, 'A century's change: natural to cultural vegetation in New Zealand', was more ambitious and appeared in the *Review* a year later, drawing directly on Sauer (Cumberland 1941).

Cumberland was also exploring the much wetter hill ranges of the North Island. He spent his first New Zealand summer, 1938–9, on a farm managed by the brother of his future wife, Marjorie, 40 km inland from Wanganui. Cumberland had met Marjorie ten years earlier in Bradford; in 1940 she joined him in New Zealand and they were married on 30 November. They spent their honeymoon in a borrowed car travelling through the remote central North Island ranges, making field notes as they went. Much of this country was recently deforested and severely eroded, as well as tectonically active. Their route further fuelled his interest in soil erosion. This had been ignited by the 1938–9 visit, and had already prompted him to work with Lance McCaskill in Christchurch to persuade government, successfully, to modify its rivers control bill as the Soil Conservation and Rivers Control Act of 1941 (McCaskill 1973; O'Connor 1980).

The Act led to the establishment of the Soil Conservation and Rivers Control Council (SCRCC). In early 1943, the Council assisted Cumberland, by arrangement with the Department of Public Works, with more concerted fieldwork by land and by air. This provided the material for his DSc thesis (1943b) and subsequent research monograph, published by the Council, on soil erosion in New Zealand (1944a). Inspired by a Sauerian concern for the destruction of resources, both works were also organized on clear regional principles. This was in keeping with the pattern of the catchment boards being set up alongside the SCRCC. One of the first was the North Canterbury Catchment Board, of which Jobberns was an elected member. Cumberland was appointed as its part-time soil conservator, the first in New Zealand. He served two summers in this role, working on land use and soil erosion surveys.

In early 1946 the Cumberland family (there were now three children) moved north to Auckland following his appointment as Senior Lecturer-in-charge of the new Department of Geography at Auckland University College. In its first year, the department – as a one-man operation – had the largest Stage 1 intake on campus. As at Canterbury, many students were either part-time, being teachers anxious to gain qualifications in a subject for which they were already responsible, or returned servicemen. Most classes therefore had to be held after 4pm; facilities were poor and resources in short supply. But in 1947, two new staff members were appointed: Ron Lister, who had been a student at University College, London in Cumberland's time, and Jim Fox, his friend from Nottingham. Fox became deputy editor of the *New Zealand Geographer*, the journal that Cumberland had established in 1945 for the New Zealand Geographical Society, which he (Cumberland) had also helped to set up.

Cumberland was appointed to the new Professorship of Geography at Auckland in 1949. By 1955, there were eight staff members. He continued as an active researcher into the 1960s, and encouraged many of his staff and students in this respect (Anderson et al. 1996). He steered the department towards an engagement with research in the islands of the south Pacific, following a trip he undertook to Fiji, Tonga, Samoa and Rarotonga in 1948. In the early 1950s, he secured Carnegie funding for a departmental research project on Western Samoa, later edited as a book (Fox and Cumberland 1962). His book *Southwest Pacific*, on Australia, New Zealand and the Pacific islands, was first published in 1954. He maintained this level of activity despite the demands on his time of his election to Auckland City Council in 1953. He was drawn into local body politics by his commitment to the practical applications of geography, one that he had emphasized in his inaugural address as a new staff member (Cumberland 1946a). He was appointed Chair of the Town Planning Committee and, as council's representative on the Auckland Regional Planning Authority, also became its Chair.

Cumberland was by this time well known internationally. On his first study leave outside New Zealand, he was offered a visiting professorship at Berkeley with Carl Sauer for the fall semester 1950, and another with Hartshorne at Wisconsin in spring 1951. In the event, he was able to accept only the second. But it was on this leave that he was also offered, without having made application, the chairs at both Toronto (previously held by Griffith Taylor (*Geographers* Vol. 3) and Manchester (earlier occupied by H. J. Fleure (*Geographers* Vol. 11). He and Marjorie decided to remain in New Zealand, primarily for the benefit of their children's education. In 1960, he was nominated by Dudley Stamp as one of the Vice-Presidents of the International Geographical Union (Cumberland 2011, 246), and served two four-year terms. In 1962, however, he stood down from the city council in view of his wife's ill health. In June 1963, she died unexpectedly. His research career never recovered momentum, although he published a series of short 'topical geographies' for schools in the second half of the 1960s. He also wrote short articles in trade journals, mostly about farming, and became a well-known commentator and speaker in this field.

Nonetheless, the initiative for which Kenneth Cumberland was to become a household name still lay ahead. In 1978, he retired from the University of Auckland to set about making a television series, *Landmarks*, in ten 1-hour episodes, a role for which he was sought out. The programmes encapsulated his career long interests in the development of New Zealand's cultural landscapes, and the opportunities and problems arising from their rapid transformation. They took four years to make, supported appropriately by the Department of Education, for showing in schools as well as on prime time television. *Landmarks* was the most ambitious series to have been produced in New Zealand, and was consciously conceived by the producer, George Andrews, in the spirit of the epics of the day, Alastair Cooke's *America* and Kenneth Clarke's *Civilisation* (Pawson 2007, 2011).

There are several reasons why *Landmarks* was memorable. Cumberland fronted it to camera, using many of the locations that had featured in his earlier research career. His rich Yorkshire voice carried the resonance of 40 years of public speaking and radio talks. And he wore the mantle of an experienced farmer, having first bought land on the hills south of Auckland in 1964, after the death of his wife. Over the years he accumulated significant land holdings in partnership with his son. He called his first farm Kettlewelldale, and his last independent home, on his younger daughter's farm, Kettlewell Cottage. He twice re-married. He lived to the age of 97, and died in Auckland in 2011.

Scientific Ideas and Geographical Thought

Cumberland's name is most readily associated in New Zealand with 'areal differentiation' or 'Hartshornian orthodoxy' (Moran 2000; Bedford 2011). His intellectual influences were however much wider, and he was to claim on retirement that what he called the 'chorological approach' was but one way of providing a coherent and defensible intellectual position. It was the ability of both students and researchers to articulate such a position that he considered necessary – even if none was left in doubt as to his own commitment to Hartshorne (Moran 2000). In turn, the importance of research was one of his central beliefs. This was not something taken for granted in the late 1930s. In fact Karl Popper, his mentor at Canterbury, had been told on arrival 'that I should be well advised not to publish anything while in New Zealand, and that any time spent on research was a theft from the working time as a lecturer for which I was being paid' (1976, 119).

In 1945, the year that Popper left Christchurch, a group of six senior academics published a short pamphlet, *Research and the University*. It was based on ideas that Popper 'poured forth', which were then taken apart by two chemists with the pieces being assembled by a geologist (Gardner et al. 1973, 308). The authors stated that 'We do not accept the point of view that teaching is the main function of the University', instead arguing that teaching and research 'should be co-ordinated and combined'. '[I]t is essential', they wrote, 'that the research function of the University teacher should be fully recognised', and that this function should be enabled within 'the spirit of free inquiry' (Allan et al. 1945, 1–2). Cumberland had taken such ideas to heart from the start of his career in New Zealand, encouraging student research in the field, and devoting himself to lengthy periods of summer fieldwork to inform his publications. In this respect he was much more active than his other Canterbury mentor, George Jobberns.

Apart from Popper and Jobberns, his main influences were undoubtedly Hartshorne and Carl Sauer and, as a student, George Stapledon. Hartshorne shaped Cumberland's thinking in favour of regional geography, although Sauer, from whom he drew his interest in landscape development, also recognized the importance of areal study (Mayhew 2011). In retirement, Cumberland was to write that his 'major interest in geography [had been] the transformation of the New Zealand landscape' (Cumberland 2011, 141). Sauer's focus on the role of human agency was of huge significance to him, whereas Stapledon also inspired the concern with the wise use of land, agricultural potential and development. This linked to Cumberland's belief in the importance of applied geography, and in making geographical insights accessible for the future and to as many audiences as possible. In his inaugural address at Auckland, he spoke of the need 'to maintain and pursue close, constant and vital relations with the community' (Cumberland 1946a, 20).

Many of these themes are anticipated in the paper of which he later said he was most proud (Cumberland 2011, 141). 'A century's change: natural to cultural vegetation in New Zealand', was published in the *Geographical Review* in 1941. It is an overview of the country's resource development remarkable for its breadth from a recently arrived immigrant. It is written in a characteristic style: confident, lucid and insightful. Its third line, referring to the speed of landscape change following European settlement of New Zealand announces: 'What in Europe took twenty centuries, and in North America four, has been accomplished in New Zealand in a single century – little more than one full lifetime'. This sentence has been often quoted, but not so much the two that follow: 'Although extensive and achieved with remarkable speed, this accomplishment can scarcely be regarded as permanent. Destruction of natural vegetation, animal life, and soil properties has been widespread, and nature's revenge is already in evidence – and at an ever accelerating rate' (Cumberland 1941, 529).

This article draws directly on Carl Sauer, whose presidential address to the Eighth Annual Meeting of the Social Science Research Conference of the Pacific Coast is cited on the opening page. The parallels are clear. Sauer wrote 'In social science interpreted as culture history there is a dominant geographic theme which deals with the growing mastery of man over his environment. Antiphonal thereto is the revenge of an outraged nature on man. It is possible to sketch the dynamics of human history in terms of this antithesis' (Sauer 1938, 765). A key element of Sauer's address was the significance of human agency compared to physical causes of change. Sauer traced this back as far as the Neolithic, but focused on the period from the late eighteenth century. He identified three sets of effects: species extinction; the restriction in range of 'useful species', and what he called 'the most widespread and most serious debit to be entered against colonial commercial exploitation': soil erosion (Sauer 1938, 771).

The understanding of these effects was to be central to Cumberland's research and writing. The 1941 article placed due emphasis on 'the impact of Europe', using five headings: 'Golden fleece and bonanza wheat', 'Toll of gold mining', 'Timber and gum exploitation', 'The dairying industry' and 'The pioneer fringes'. These successive human impacts are drawn as particular to specific places. Here the hand of Hartshorne is also in evidence: 'The ousting of the indigenous plant cover and the partial establishment of an exotic (mainly European) vegetation in its place is at one and the same time the essential theme of the history of New Zealand and the foundation of *the regional differentiation of its area*' [emphasis added] (Cumberland 1941, 529). Two maps accompanied the article, both compiled by Cumberland, who was a skilled draughtsman, from the acknowledged work of others. That of 'primitive vegetation' in 1840 was contrasted with 'cultural vegetation' in 1940. Although unstated, the second date was the New Zealand Centennial, the anniversary of the signing of the Treaty of Waitangi between Maori tribes and the British Crown. But the analysis does not end in 1940; rather it sketches an argument about agricultural diversification that Cumberland was to revisit in greater depth and with increasing frequency in later years.

It was the delineation and understanding of soil erosion that was to claim his attention next, however, and about which he was to make his name. There had been a smattering of brief official publications on the topic in the late 1930s, but nothing substantial. Cumberland's observations in the Wanganui hill country in the summer of 1938–9 sparked his interest; the field experience on his honeymoon in December 1940 extended it. Subsequent, more comprehensive, investigations were assisted logistically by the Public Works Department. The resulting DSc thesis is dated October 1943; the book that followed was published by the SCRCC in 1944. Both bore the subtitle

'A geographic reconnaissance'. The work was intended as an overview, and as a geographical not soil science analysis. 'In what follows', he wrote, with the emphasis in the original, 'the thesis is advanced that:

> *soil erosion . . . is a distinct and separate problem in different areas; that it finds differential expression in terms of the severity with which it occurs and the forms it assumes in the landscape of different regions; that each region recognised on these bases requires different solutions of its own individual soil erosion problems, and finally, that soil erosion must be approached regionally and in full knowledge of the intricate interrelation of physical and cultural conditions under which soil wastage has been hastened in differentiated regions.* (Cumberland 1944a, 7–8)

The influence of Hartshorne is apparent in this approach, but it was Sauer that he referenced as a source legitimating 'regional synthesis' as the basis of 'scientific investigation of land use and soil erosion problems and in the planning of future land policy' (Cumberland 1944a, 8).

Although only 500 copies of the book were produced (and the same number again in a reprint in 1947), it received considerable notice in the newspapers in each of New Zealand's four main cities, including in a number of leader columns. Echoing the tone of an article that Cumberland had written for the New Zealand Army Education Branch in 1943, the *New Zealand Herald* in Auckland began: 'As he turns over page after page of photographs . . . the New Zealander who loves his country will hardly escape feelings rather like those of one who peruses an illustrated official report on wartime atrocities.' The use of such metaphors was prominent in the discourse of soil erosion at the time (Roche 1997). 'Terse and arresting', said the *New Zealand Freelance*; and by far the longest notice, in the *New Zealand Listener*, was headed 'Sleepers Awake! Geographer Sounds An Alarm'.

There were also a number of positive academic reviews. Cumberland's London mentor, R. O. Buchanan, in the *New Zealand Geographer*, described the book's appearance as marking 'a stage, almost an epoch, in the development of geography in New Zealand' (1946, 272). Graham Lawton, in the *Geographical Review*, gave an affirmative summary. In *Nature,* a leading imperial authority on the topic noted that 'the author is more concerned with soil erosion than soil conservation', but urged that it 'be read by everyone who is interested in this grave and widespread disease of the land' (Jacks 1945, 703). In 1947, it was included in a list of 'Classics in the soil conservation field' in the *Journal of Soil and Water Conservation* compiled from conversations with 'experienced soil conservationists' by the head of the Pacific Region of the US Soil Conservation Service. He commented that the book was little known in the United States but that it 'is probably the best survey report ever published on the land situation of any country' (Van Dersal 1947, 110).

This book, and associated research such as an article in the *New Zealand Geographer* in 1945 that assessed farmers' burning practices in the South Island high country, has maintained a profile ever since. New historical work (Peden 2011) has criticized Cumberland's use of material from earlier writers, for implying too much of a connection between wanton burning and soil erosion in the nineteenth century, when burning practices can be shown to be purposive, as indeed Cumberland himself showed that they were by the 1940s. Research by mountain lands scientists has argued that more emphasis should be placed on the ongoing dynamics of physical change in tectonically active environments (Mather 1982; McSaveney and Whitehouse 1989). O'Connor, however, wrote that 'We served ill the legacy of Cumberland when we stressed the summary element in his findings and neglected the detail in his observations that suggested more assiduous examination and measurement by those that followed him' (O'Connor 1980, 199).

Those that openly neglected to do this, denying the existence of a soil erosion problem, came in for trenchant criticism from Cumberland. This was the fate of the report of the Royal Commission on Sheep Farming in 1949. Writing in the *New Zealand Herald*, he said: 'There can be no report that is at the same time based on such false and confused premises, so self-contradictory or so badly written as to become at times incomprehensible as this latest product of the Government Printer' (Cumberland 1949a, 9). The prompt for this vehemence was the Commission's failure in his view to distinguish between 'erosion' as a physical process and 'soil erosion' as a humanly accelerated phenomenon, thereby absolving the 2,000 or so farmer witnesses of any responsibility. There were to be many examples of Cumberland taking clear stands in his career. Another was in the early 1960s, in Jobberns' festschrift, when he presented a vigorous and lengthy case that pre-1840 deforestation in New Zealand was largely the product of Polynesian, that is, Maori, agency in the landscape, not climate warming and dessication around 1200 AD as an emerging orthodoxy was suggesting (Cumberland 1962a).

Interestingly, this position undermined his own assertion in the 1941 paper that Maori were 'essentially conservational' (Cumberland 1941, 531). In 1961, he argued in a paper again clearly inspired by Sauer that 'The Maori had a greater role in changing the face of the land than has hitherto been apparent; and the evidence suggests that fire was his most important weapon' (Cumberland 1961, 145). This view has been essentially vindicated by recent research (McWethy et al. 2010). Much of Cumberland's work in historical geography was however more conventional, echoing the cross-sectional narratives associated with Clifford Darby (*Geographers* Vol. 26) in Britain (Darby 1936). These were consistent with the chorological approach that Hartshorne (1939) was to expound. Cumberland produced five synchronic cross-sections of New Zealand (for 1250, 1780, 1838, 1853 and 1881), one being jointly authored. They appeared from 1949 to 1962, when his stated view was that 'changes from time to time are the concern of history; differences from place to place are the concern of geography' (Cumberland 1955a, 185). But as a later review of historical geography in New Zealand pointed out, this had not 'prevented Cumberland from writing another kind of historical geography', concerned with the evolution of the cultural landscape (Perry 1969, 94–5).

There is more than one answer to this apparent paradox. Cumberland considered it necessary to assert a chorological approach for both pedagogical and, at various times, for political reasons (Moran 2000; Pawson 2011). But even the cross-sectional studies were 'commonly . . . concerned with longer periods of time than their titles might suggest. The forces and characteristics of change – past, present, and future – cannot be ignored' (Perry 1969, 98). In such ways, the influence of Sauer was making itself felt, an approach that had enabled Cumberland to tackle a number of applied issues in a series of papers in the mid-1940s. These included a method of regional land inventory (1944c), echoing the initiative of Stamp and Willatts in the United Kingdom, but adapted from work by the Tennessee Valley Authority. Its purpose was to provide means of assessing the extent of 'physical land crises' (1946c, 137), including the spread of noxious weeds and accelerated erosion. Next, he argued for 'an overall forward-looking agricultural plan that insists primarily on giving the land greater care and attention' (1947, 191). This was to put geography firmly at the service of applied issues, which the soil erosion book had started to do so well. It was also to emphasize the primacy of human agency and to dispel once and for all what Cumberland regarded as the baleful legacy of environmental determinism (Cumberland 1946a; Roche 2011).

In his inaugural address in Auckland, he railed against the determinist 'view of geography which today still has widest currency in New Zealand schools, just where it is most dangerous' (Cumberland 1946a, 11). Discussing the role of human agency in the region in which he had recently lived, he said: 'Physical factors have contributed little to these frequent transformations in the life and landscape of the

Canterbury Plains' (Cumberland 1946a, 13). Instead, he stated, 'geography seeks to discover "what places are like"' (Cumberland 1946a, 8), a statement repeated in his presidential address to the first New Zealand Geography Conference, held in Auckland in 1955 (Cumberland 1956b, 3). This rather straightforward assertion of a Hartshornian viewpoint was one that Cumberland regarded as giving geography, still needing to establish and differentiate itself among subjects in the New Zealand university and school system, the sort of intellectual standing that Popper, who had read Cumberland's copy of Hartshorne, assured him could be found in the 'chorological' approach (Cumberland 2011, 114).

New Zealand: A Regional View was authored by Cumberland and Fox and first appeared in 1958. It followed a successful series of regional studies that the Department of Education had published in its *Post-Primary School Bulletins* between 1950 and 1953. These had been designed and introduced by Cumberland, who wrote a number of them, and coordinated other geographers as authors of the rest. The book was new, in as much as each chapter was rewritten by Cumberland and Fox, 'not without controversy among the authors of the original individual bulletins' (Moran 2000, 432). They also added two additional chapters: an overview of the country as a whole, and one on the four main metropolitan centres. The opening chapter was a statement about the nature of regional geography, intended for teachers. Several reviewers admired its clarity. The *New Zealand Geographer* considered this chapter to be 'a very lucid account of the nature of regional geography' (Wilson 1959, 206); the *Australian Geographer* felt that the regional studies were 'suited to mature students as well as to secondary school pupils' (Robinson 1960, 45).

In the preface, Cumberland and Fox stressed that the book 'outlines only one view of the nature of regional geography'. As was to be expected from his statement in 1946, and in contrast to older regional texts, there was relatively little physical geography. The approach was intended to meet the authors' hope that the book 'will also prove of interest and of use to the layman' (Cumberland and Fox 1958: vi–vii). While noting the lack of footnotes and bibliography (there was a list of further readings), reviewers mostly agreed. The *Geographical Review* commented that although written for New Zealand readers, 'it will be of only slightly less interest to students of New Zealand geography the world over' (Clark 1960, 293). Leslie Symons in *Geography* observed that it 'is the only book that deals reasonably comprehensively' with the country's regional geography (Symons 1960, 145). It found a ready audience among students in courses concerning New Zealand in Britain and the United States. It went through several printings, into a new edition (in 1963) and, according to Cumberland's own meticulous records, sold nearly 61,000 copies in the years to 1972.

Even so, this book probably had a less wide reach among university readers than Cumberland's slightly earlier text, *Southwest Pacific: A Geography of Australia, New Zealand and Their Pacific Island Neighbourhoods*. It shared many of the characteristics of the regional book: clear writing, lack of referencing and an energetic style. It went through four New Zealand editions to 1968, but was also published under the imprint of McGraw-Hill in New York (1955, 1958), Praeger, New York (1968), Methuen, London (1955, 1958 and 1968) and Asakura, Tokyo (1971). Sales amounted to over 41,000 until 1975. This book was even more widely reviewed in overseas journals, including short notices in *Erdkunde* and *Annales de Géographie*, and longer ones in English language titles. Reviews and articles also appeared in many New Zealand newspapers, just as had been the case for *Soil Erosion in New Zealand* a decade earlier. It was widely praised for its accessibility and interest, and for its provision, in the wake of World War II, of useful information about New Zealand's Pacific neighbourhood. Cumberland was clear about designing a book 'to make available to the people of Australia and New Zealand especially, and to the younger generation of future politicians, administrators and businessmen in particular, an objective geographic account of the territories of the southwest Pacific' (Cumberland 1954a, vi–vii). Andrew Clark, reviewing the book in the *Geographical Review*, wrote that if it did not achieve its aims 'it will be because the readers to whom the book is addressed fail to get it into their hands. Few who dip into it will want to set it down until they have sampled it generously – perhaps as remarkable a statement as one could make about a work designed as an elementary text' (Clark 1956, 294).

Cumberland's interest in the Pacific Islands began with his research visit in 1948 during which he recorded what he experienced in a field notebook complete with sketch maps and diagrams. He then published a paper that included the subheading 'Forms of traditional land ownership hamper progress' (Cumberland 1949c, 7), a view based on the urgency of food production. Another, on the value of land use surveys (Cumberland 1951a), stemmed from his earlier paper on the unit area method of land inventory (Cumberland 1944c). A Carnegie grant (1954–7), and later smaller grants, enabled the Auckland department, including postgraduate students, to work in Western Samoa in the lead-up to independence (Anderson et al. 1996). The resulting book represented the culmination of Cumberland's interest in land use survey and development (Fox and Cumberland 1962). It was criticized at the time for inadequate description of the structure and functioning of the land tenure system and its main conclusion that this had to be changed if food production was to be increased for a rapidly growing population (Bedford 2011) – not a view that he seems ever to have advanced about Maori land tenure in New Zealand. A much later assessment of his Pacific interests concluded that 'One of Cumberland's major legacies was the successive generations of graduates in the 1950s and 1960s that he and his staff encouraged to engage with issues related to the contemporary geography of New Zealand's Pacific neighbourhood' (Bedford 2011, 135).

Influence and Spread of Ideas

Cumberland's wider influence reflected his vision of geography as a discipline for developing a well-informed citizenry, whether at school or university level, or among a wider public. He made few published methodological statements, but delivered two at key points in his career. The first was his inaugural address at Auckland (Cumberland 1946a), the second his presidential address to the first New Zealand Geography Conference (Cumberland 1956b). In the former, he was seeking to establish a discipline in a university that had no history of teaching it in a form then recognizable. In the latter, he was both promoting and seeking to protect this vision in a contest that was emerging with the geographers at Victoria University of Wellington. He pointed to 'the geographer's "pragmatic sanction" as both fulfilling a people's curiosity about the places that comprised their world and securing "practical advantages" from that knowledge' (Wilson and Henry 2011, 121).

These positions were advocated in his influential school texts. In 1943, the first edition of *Whitcombe's Map Reading in Geography* appeared. It carried the subtitle *The Interpretation of Topographic Maps and Geographic Photographs.* In the Foreword, Cumberland wrote:

> If the function of geography in schools is to equip the citizens of the future to think sanely about the political and social problems in their own district and nation and in the world at large, then an understanding of maps and ability to use them well will certainly prove a great advantage. With this in view, map reading should begin where geography begins – in the primary school. There should be topographic maps in every school; in every home. (Cumberland, 1943a, 1)

The problem was that topographic map making was in its infancy in New Zealand, and no sheets were available until after World War II (Cumberland 1946b). The book was widely used, going through eight editions before 1968.

In November 1940, Cumberland had signed a contract to produce an atlas for use in New Zealand and Australian schools, with the Christchurch firm that was to publish most of his books, Whitcombe and Tombs. This contract stated that the author was to receive a 10 per cent royalty but only after the first 5,000 copies had been sold. Cumberland (2011, 125) later recorded that Bernie Whitcombe 'didn't realise that he had been dealing with a Yorkshireman', although advice from Jobberns who had learned the hard way how to deal with the publisher is likely to have been important (Roche 2011). The atlas went on to sell nearly half a million copies over its 13 editions, still being in print in the 1960s. Cumberland drew the maps, and thanked his colleague Fox for 'assistance in the work of revision' in later editions. In 1949 he added *This is New Zealand: A Pictorial Geography*, to match 'the recent insistence upon the employment of visual aids in education'. It was a thematic geography with photographs. The preface to the first edition (this too went through several editions, including one from Methuen in London in 1953) highlighted an additional purpose:

> It should provide for people overseas a picture of New Zealand that is attractive, concise, and well-balanced yet neither coloured by propaganda nor embellished with the exotic and bizarre as is so much formal publicity. (Cumberland 1949e, 3)

It sold over 73,000 copies, producing royalty income well into the 1970s.

His commitment to producing such books reflected not only the need to provide resources for high-school teaching, but also the sort of materials that he believed would awaken the fascination for New Zealand's places and landscapes that he had shown in his 1940 and 1941 *Geographical Review* papers. These had been lavishly illustrated with photographs. In *A Pictorial Geography* he wrote: 'Apart from the map in its many and varied forms, perhaps the most valuable aid in the teaching of regional geography is the ordinary photograph, judiciously selected for its objective description of fact, its human interest and its general appeal to the pupil' (Cumberland 1949e, 3).

The regional geographies in the *Post-Primary School Bulletins* for the Department of Education, which Cumberland wrote, or oversaw, followed. An early one was a statement on the regional geography approach. It contained, however, 'a word of warning': 'Neither a topographic map or a photograph will tell you *why*. The map may supply hints – it cannot supply the complete answer to a "why" question' (Cumberland 1950c, 175).

These articles were later to be reworked as *New Zealand: A Regional View* (Cumberland and Fox 1958). Together they formed the basis of secondary school teaching for more than 20 years (Moran 2000). The material was often felt, however, to be too advanced for its purpose (Roche 2011). The reviewer of the book in the *New Zealand Geographer* observed that when 'the original bulletins first appeared, there was general regret amongst teachers that their expression was often too difficult for many secondary school pupils'. He noted that the book 'is simpler and better adapted to the calibre of most of its readers [even if] sometimes an almost Johnsonian broadside such as ". . . the elongated and attenuated shape of Eastland" contrasts sharply with the simplicity of "Westland is long and narrow"' (Wilson 1959, 207).

There was no doubt however that Cumberland's passion for geography was widely appreciated. Reporting on the first New Zealand Geography Conference which, then as since, included a mix of academic and school attendees, one said that 'geographers are keen, fit, and enthusiastic, and it's catching' (Downs 1955, 11). Another, commenting on 'remarkable changes in education', and in particular in geography, in the preceding ten years, wrote this 'has been in very large measure due to the influence and enthusiasm of Professor K. B. Cumberland . . . The vitality and zeal which he has transmitted to the growing ranks of "new" geographers was reflected in the enthusiastic response accorded to his presidential address' (McNaughton 1955, 298). The '"new" geographers' were those teaching the regional geography of the *Bulletins* in schools. Yet in this report, the author also noted that a 'disturbing' feature was 'an over-anxiety of geographers to justify their position, possibly the understandable sensitivity of the converts of a new religion . . . [One] geographer was criticized by his fellows for daring to preach something other than the party line' (McNaughton 1955, 298).

This was a reference to the debate about the future of New Zealand university geography between regional orthodoxy and a perspective from the department at Victoria University College in Wellington that drew more widely on the social sciences (Moran 2000; Bedford 2011; Wilson and Henry 2011). The chorological view espoused by Cumberland at Auckland was also practiced at Canterbury and Otago. There had been movement of junior staff from Auckland to Canterbury, and Ron Lister had assumed charge at Otago in 1951. But the two leading geographers at Victoria, Harvey Franklin and Keith Buchanan, were both appointed from Britain (in 1951 and 1953 respectively) and trained in the social sciences. It was Franklin who at the conference was criticized for arguing 'the necessity on the part of geographers for a closer acquaintance with the ideas, attitudes and work of the social and cultural sciences' (1955, 1). Buchanan was particularly interested in the socialist world and in his inaugural address (1954) had approvingly quoted Sauer's argument that a discipline defined by defence of its borders was 'likely to face extinction' (Sauer 1941, 4). Cumberland, however, saw such views as 'deep cleavages' that he considered would not 'find favour, respect and acceptance amongst the other disciplines' (1956b, 2).

In this 'celebrated debate' (Bedford 2011, 129), Cumberland, having established so many of the disciplinary spaces within which New Zealand geography operated at the time, had a great range of performative practices at his disposal (Pawson 2011). He ensured that Hartshorne was compulsory reading for all Masters students at Auckland (Moran 2000). His influence on school curricula and through books was extended by the *New Zealand Geographer*, which he set up in 1945, and edited until 1955, apart from his sabbatical year in 1951, when Fox took over. Fox again assumed the reins in 1955, through to 1962 (Wilson and Henry 2011). The Victoria geographers set up their own journal, *Pacific Viewpoint* (later *Asia-Pacific Viewpoint*) in 1960, as an outlet for a geography framed by issues of underdevelopment: a quite different approach to the modernization thesis explicit in Fox and Cumberland (1962). This was the context for the remark in the preface to the second edition of *New Zealand: A Regional View*, that:

> Geography, whether conceived as a study of relationships between man and the earth or as a study of what places are like, is inevitably concerned with the swiftly changing world scene. Nor is this dynamism confined to socialist and planned societies. Change, development, growth and progress are characteristics also of the New Zealand scene. (Cumberland and Fox 1963, 7–8)

Buchanan was critical of the adequacy of 'areal differentiation' for understanding the dynamism of socialist societies (Buchanan 1961; Pawson 2011).

There were essential aspects of Cumberland's commitment to the New Zealand scene that have left a less obvious archival trace than research and publications. The first of these was his teaching. In the classroom, he was a dominant presence (Moran 2000). Norman Whatman, later to be a Teachers' College Vice-Principal for many years, remembers Cumberland's 'quiet purposeful entry into the lecture room' in the first term of 1938. 'He was appalled at our generally casual approach to the discipline of academic study . . . I wonder now whether the "culture shock" for him was as great as the "academic shock" we received?' (in Macaulay 1987, 20). For his part, commenting on the Stage 1 geography examination at the end of 1939, Cumberland wrote:

> The scripts showed a wide range of ability but the large majority revealed a surprising immaturity of judgment, a lack of basic factual knowledge, serious irrelevancy and frequently confused and muddled thinking. The immaturity and lack of basal knowledge are no doubt traceable to the limited extent and low average standard of the teaching of geography in schools. (in McCulloch 1992, 170–1)

Student reminiscences from this period recall Cumberland's 'mind-expanding lectures on Europe' (McCaskill, M., in Macaulay 1987, 49). 'His depth of knowledge and ability to organise his material made him an excellent lecturer. His dry sense of humour perhaps escaped some of his students' (Barley, A., in Macaulay 1987, 33). Fieldwork with Jobberns and Cumberland is often remembered, a tradition that Cumberland took with him to Auckland, and which has persisted, albeit in re-invented form, till the present (Winder and Lewis 2010). There, as at Canterbury, the large majority of students were teachers studying part-time for degrees, or intending teachers. Cumberland, given his views on the wider roles of geography, complained that 'young fully-trained geographers can find few openings other than in the pedagogical field' (Cumberland 1951b, 176). Their education was not focused on regional geography alone, however, and from the late 1940s through to the early 1960s, 'the undergraduate programme comprised three papers that were *systematic* and three *regional*, with associated practical courses' (Anderson et al. 1996, 17, emphasis in original).

The Auckland department initially grew rapidly under Cumberland's tutelage. Students and lecturers from the department contributed to the flow of well-qualified people as demand for university staff expanded in New Zealand and around the world. Les Curry, who was appointed to the staff in 1953 from Durham, and who was awarded Auckland's first PhD in Geography in 1958, left the next year for North America, spending most of his career as a leading theoretician at the University of Toronto. In 1955, Gordon Lewthwaite, one of the department's first Masters students, returned to the staff having undertaken a PhD at Wisconsin; he resigned four years later to take up a position at California State University (Northridge). In 1960, Bryan Farrell, who had been on the staff since 1951, and gained his PhD in the department, moved on to a distinguished career in North America. Another appointment from this period was Gerard Ward, who was to become chair of the Research School of Pacific Studies at the Australian National University.

With increasing numbers of academic colleagues, but a decline in student enrolments after 1952 as the post-war surge passed through (as well as the growing availability of new majoring subjects), Cumberland found space to take up a new interest. This was indirectly foreshadowed in his 1946 address when, in a section headed 'Geography and government', he said it is 'essential that all of us should appreciate the problems of our own and other localities, otherwise democratic government cannot function to the full' (1946a, 23). He was elected to the Auckland City Council, and was to serve from 1953 to 1962. This carried with it parallel memberships of the Auckland Metropolitan Drainage Board and the Auckland Regional Planning Authority, the second of which he chaired between 1956 and 1959 (Cumberland, 1971a). As chair of the Council's Town Planning Committee, he was also responsible for directing preparation of the city's new planning scheme.

Cumberland was enticed to stand for the council when a new political party, Civic Reform, was formed to oppose an open sea sewerage disposal upgrade for Auckland. After the 1953 election, this was successfully overturned, and replaced with a more sophisticated treatment system. Cumberland describes the outcome in the book that accompanied *Landmarks* (Cumberland 1981, 261–4), although he does not comment on the part he played, which is detailed in his memoir (Cumberland 2011). Nor does *Landmarks* reveal his role in urging the development of a master transport plan for Auckland that would include both the electrification and partial undergrounding of its suburban railways, as well as a new motorway framework that he saw as necessary to ease road congestion. Only the motorway system was funded over succeeding decades, but for long without provision of cross-city links, thereby adding to the problem it set out to resolve. In 1957, he was urged to stand for the Mayoralty, but was refused permission by the College authorities. 'Their interdiction was foreseeable, but was all the same for the City a matter of regret' (Bush 1971, 505).

In 1962, Cumberland, who had polled second among 60 candidates at the council elections three years earlier (*New Zealand Herald*, 13 August 1962), decided not to stand again, due to his wife's ill health. His research career also slowed about this time: he produced few papers of academic substance after 1963. His time was increasingly taken up with overseas travel associated with his IGU responsibilities, and with new pursuits, notably farming, and as a commentator on both agriculture and everyday affairs. He had been a regular radio broadcaster back to his earliest days in Canterbury and in Auckland, often in conjunction with other departmental members, and occasionally for the BBC. He also wrote frequently for newspapers. In the 1960s, he was a well-known contributor to *Looking at ourselves*, a 15-minute programme that discussed current affairs on the YA stations in the four main urban centres. Both in the media, and in talks given up and down the country to farm industry groups in the 1960s and 1970s, he was critical of government policy, as he had been much earlier (Cumberland 1947). One extract is typical of his views: 'It's indicative of our wooly thinking, and of attitudes ingrained since the 1930s, that the National Development Conference so presented its targets as to lead the average [reader] to believe that New Zealand's future lies in manufacturing and that farming has had it. This is not only unfortunate. It could be psychologically and economically disastrous' (*Looking at ourselves*, 14 September 1968).

He took up this theme on the second occasion on which he was president of a New Zealand Geography Conference, in 1967 (Cumberland 1968). His argument, as anticipated in the 1941 paper in the *Geographical Review*, was that agricultural intensification and diversification into new markets and products, including horticulture, were the answers to maintaining national prosperity, rather than trying to foster small-scale industry behind tariff walls. He elaborated it over many years (Cumberland 1979). In line with this, given the adoption of the motorway plan for Auckland during his years on the city council, he shifted his view on the need to contain urban sprawl (Cumberland 1955b, 96) towards arguing that the 'far-flung urban-rural fringe of low density settlement . . . will at least provide the more imaginative regional planners with exciting opportunities for experiment and innovation in physical planning and land use design, nicely mixing intensive horticulture, extensive residential uses, reserves, open spaces, recreational facilities' (Cumberland 1971a, 19).

He took his own advice, first buying land half an hour's drive beyond the end of the Auckland southern motorway in 1964. Other farms were purchased, both for sheep and dairy production; the first was later subdivided for rural housing, and land acquired at Te Kuiti in the King Country for the '"Carpetmaster", a breed we have evolved ourselves for carpet wools' (in Ray 1981, 15). For some years, he advocated 'forest farming', or 'three tier farming', in which trees, pasture and horticultural crops such as tamarillos could be grown together (Cumberland 1976). Politically, his interventions were not always appreciated. He thanked the organizers of a lecture that he delivered in 1971 for extending the invitation to speak as it had helped 'to restore my confidence' because in discussing recently the future of New Zealand agriculture,

I have been singled out by the heavy artillery of at least four Ministers of the Crown and referred to as either '"one of the less well informed", or as doing this country "a disservice", or, more bluntly, as a "saboteur"' (Cumberland 1971a, 1). For someone so long regarded as the heavy artillery of his first profession, academic geography, this was indeed a change of places.

Conclusion

In 1951, John Kirtland Wright, Secretary of the American Geographical Society, constructed a 'celestial hierarchy' based on contributions per author to the Society's journals since 1900. Kenneth Cumberland's six papers in the *Geographical Review* placed him at the rank of 'Archangel', even though the first did not appear until 40 years after the start of Wright's census (Wright 1952, 294). Cumberland's standing was further recognized by the award of both Fulbright and Carnegie Fellowships in the 1950s. His election to the Executive of the International Geographical Union for two terms in 1960 was as Vice-President representing the interests of Britain and Commonwealth countries (Lister 1978). In 1973, he was the first professional geographer to be elected a Fellow of the Royal Society of New Zealand, and was later appointed CBE, Commander of the Order of the British Empire.

In his letter of application to Canterbury University College in 1938, Cumberland already showed the high expectations he had of himself: 'I have especially in mind the future of my subject in the Dominion. I should consider it, not only a very great honour, but also a formidable task calling for exceptional effort, to assist in laying the foundations of Geography as a subject in the University of New Zealand' (in Macaulay 1987, 24). In training generations of teachers to degree level, and in writing much of the textbook material that they used in schools up until the 1960s, his impact was to reach far beyond the universities. His concern that curiosity about landscape should take geography into the public realm was also prosecuted to great effect, with the reception afforded his views in newspapers, on radio and eventually on television being considerable and long lasting, from the 1940s to the 1980s. The applied aspects of his work, particularly as they affected land use and farming, were well covered in the festschrift presented to him by colleagues from around the country on his retirement (Anderson 1980). His major academic contributions are however better represented by his journal articles, his soil erosion book and, as an overview, the *Landmarks* programmes.

Even if Cumberland's intellectual attachment to Hartshorne seemed to some to leave little room for alternative readings of geography, his influences were more catholic, the Auckland department more widely rooted, and some of his own best work informed by other thinkers, notably Sauer. He introduced and fostered the practice of research-based inquiry into New Zealand geography, leading not only by example, but also, with the Western Samoa study, encouraging cooperative research as 'a distinct and early break with the "lone scholar" tradition' (Anderson et al. 1996, 23). Murray McCaskill wrote of Cumberland's profile as a public figure that 'Few geographers in any country can have held for ten successive weekly sessions a television audience estimated at up to one-third of the national population, then capped this success with sales of tens of thousands of copies of a fairly expensive book' (1983, 348). And as Cumberland himself said, reflecting later on the making of *Landmarks*, when on television 'You simply have to be bold, assertive, outspoken and single-minded' (1990, 11). It is hard to think of a better way to sum up his career as a New Zealand geographer and as a geographer of New Zealand and the southwest Pacific.

Acknowledgements

I am grateful to Tanya Cumberland, Charmaine Pountney and Garth Cumberland for their hospitality, responses to my enquiries and assistance in retrieving material from Professor Cumberland's archive (now located in The University of Auckland library). Many other colleagues have helped me with discussions about the subject, and reading of earlier drafts of this essay. They include Henry Connor, Ron Johnston, Warren Moran, Peter Perry, Geoffrey Rice, Michael Roche and Vaughan Wood.

Bibliography and Sources

1. SELECT BIBLIOGRAPHY OF WORKS BY KENNETH CUMBERLAND

1938 'Livestock distribution in Craven', *Scottish Geographical Magazine* 54, 75–93.

1940 'Canterbury landscapes: a study in New Zealand geography', *Geographical Review* 30, 19–40.

1941 'A century's change: natural to cultural vegetation in New Zealand', *Geographical Review* 31, 529–54.

1942 *Whitcombe's Atlas of Geography for New Zealand and Australian Schools.* Christchurch: Whitcombe and Tombs (published in 13 editions up to 1957; last reprinted 1966).

1943a *Whitcombe's Map Reading in Geography. The Interpretation of Topographic Maps and Geographic Photographs.* Christchurch: Whitcombe and Tombs (published in nine editions up to 1968).

1943b The regional morphology of soil erosion in New Zealand. A geographic reconnaissance, submitted in application for the doctorate degree, University of New Zealand.

1943c 'Defending our soil', *New Zealand Army Current Affairs Bulletin* 1(13).

1943d 'A geographic approach to soil erosion in New Zealand', *Australian Geographer* 4, 120–31.

1944a *Soil Erosion in New Zealand: A Geographic Reconnaissance,* Soil Conservation and Rivers Control Council, Wellington.

1944b 'Contrasting regional morphology of soil erosion in New Zealand', *Geographical Review* 34, 77–95.

1944c 'The survey and classification of land in New Zealand: a basis for planning', *Transactions of the Royal Society of New Zealand* 74, 185–95.

1944d 'High country "run". The geography of extensive pastoralism in New Zealand', *Economic Geography* 20, 204–20.

1944e 'New Zealand and the post-war world', *Agenda* 3, 33–42.

1945 'Burning tussock grassland: a geographic survey', *New Zealand Geographer* 1, 149–64.

1946a *The Geographer's Point of View. An Inaugural Lecture Delivered on 2 April 1946*. Auckland University College, Auckland.

1946b 'The status of topographic mapping in New Zealand', *Geographical Review* 36, 135–6.

1946c 'Land problems in New Zealand: a review', *Geographical Review* 36, 137–41.

1946d 'Population changes in New Zealand', *Geographical Journal* 108, 121–3.

1947 'The land and the future: a review and argument', *New Zealand Geographer* 3, 178–92.

1948 'The agricultural regions of New Zealand', *Geographical Journal* 112, 43–63.

1949a 'Views on soil erosion criticised: superficial treatment by Sheep Industry Commission alleged', *New Zealand Herald*, 18 July, 9.

1949b 'Aotearoa Maori: New Zealand about 1780', *Geographical Review* 39, 401–24.

1949c 'New Zealand's "Pacific Island neighbourhood": the postwar agricultural prospect', *New Zealand Geographer* 5, 1–18.

1949d 'Geography in the University of New Zealand', *Scottish Geographical Journal* 65, 98–9.

1949e *This is New Zealand: A Pictorial Description*. Christchurch: Whitcombe and Tombs (published in five editions up to 1960; later editions entitled *Whitcombe's New Zealand in Outline: A Pictorial Geography*).

1950a 'A land despoiled: New Zealand about 1838', *New Zealand Geographer* 6, 13–34.

1950b 'New Zealand and the southwest Pacific', *Scottish Geographical Magazine* 66, 14–17.

1950c 'Geographic regions of New Zealand: the nature of regional geography', *Post-Primary School Bulletin* 4, 161–76 (accompanied by five regional essays in this journal from 1950 to 1953).

1951a 'Geography and land-use survey in the southwest Pacific: a review and suggestion', *New Zealand Geographer* 7, 139–53.

1951b 'Worthy of his hire', *New Zealand Geographer* 7, 175–6.

1953a 'New Zealand agriculture, 1930–1950', *Geographical Review* 43, 117–19.

1953b 'Population growth in New Zealand: a review of recent census returns', *Scottish Geographical Magazine* 69, 97–105.

1954a *Southwest Pacific. A Geography of Australia, New Zealand and their Pacific Island Neighbourhoods*. Christchurch: Whitcombe and Tombs (published in four editions up to 1968, plus overseas editions).

1954b '"Jimmy Grants" and "Mihaneres": New Zealand about 1853', *Economic Geography* 30, 70–89.

1955a 'American geography: review and commentary', *New Zealand Geographer* 11, 183–94.

1955b 'Living in New Zealand cities', *Journal of the New Zealand Institute of Architects*, 22, 96.

1955c (with R. P. Hargreaves) 'Middle Island ascendant: New Zealand in 1881, Part One', *New Zealand Geographer* 11, 95–118.

1956a (with R. P. Hargreaves) 'Middle Island ascendant: New Zealand in 1881, Part Two', *New Zealand Geographer* 12, 51–74.

1956b 'Why geography?', *New Zealand Geographer* 12, 1–11.

1958 (with J. W. Fox) *New Zealand: A Regional View.* Christchurch: Whitcombe and Tombs (Second edition, 1963).

1959 'New Zealand after twenty years: a geographer's view', *Geographical Journal* 125, 38–47.

1960 'Problem and prospect in Fiji', *New Zealand Geographer* 16, 214–17.

1961 'Man *in* nature in New Zealand', *New Zealand Geographer* 17, 137–54.

1962a '"Climatic change" or cultural interference? New Zealand in Moahunter times', in M. McCaskill (ed.), *Land and Livelihood. Geographical Essays in Honour of George Jobberns.* Christchurch: New Zealand Geographical Society, 88–142.

1962b 'Moas and men: New Zealand about A.D. 1250', *Geographical Review* 52, 151–73.

1962c (with J. W. Fox) *Western Samoa. Land, Life and Agriculture in Tropical Polynesia.* Christchurch: Whitcombe and Tombs.

1962d 'The future of Polynesia', *Journal of the Polynesian Society* 71, 386–96.

1963 'Man's role in modifying island environments in the southwest Pacific: with special reference to New Zealand', in F. R. Fosberg (ed.), *Man's Place in the Island Ecosystem.* Honolulu: Bishop Museum Press, 187–205.

1965–9 *New Zealand Topical Geographies*: 17 short booklets. Christchurch: Whitcombe and Tombs.

1967a 'Neuseeland in den epochen der Moajager und Maori: ein beitrag zu prahistorischen geographie', *Die Erde* 98, 90–114.

1967b 'Geography and relations between Japan and New Zealand', *Geographical Science* 7, 1–8.

1967c 'The countryside and society', *The New Zealand Countryside in 1980. Proceedings of the Study Conference, November 1967.* Auckland: The New Zealand Institution of Engineers, 55–63.

1968 'The New Zealand economy: a geographer's view (Presidential address)', *Proceedings of the Fifth New Zealand Geography Conference.* Auckland: New Zealand Geographical Society, 7–15.

1970 (with J. S. Whitelaw) *The World's Landscapes. 5 New Zealand.* London: Longman.

1971a 'Planning a future for the land and agriculture', J. T. Stewart Lecture in Planning, No 5. Palmerston North: New Zealand Geographical Society, Manawatu Branch.

1971b *Auckland in Ferment: The Present and Future Brew.* Auckland: New Zealand Geographical Society, Auckland Branch.

1972 (edited, with photographs by W. Imber) *Neu Seeland: Antipode des Abenlandes.* Bern: Kummerley and Frey.

1973 (edited, with photographs by W. Imber) *New Zealand, Pacific Land Down Under.* Wellington: Reed.

1975 'George Jobberns', *New Zealand Geographer* 3, 1–3.

1976 'Three-tier farming in New Zealand's economic future', *Farm Forestry* 18, 37–52.

1979 'An outsider's view of the agricultural prospect', *New Zealand Agricultural Science* 13, 18–25.

1981 *Landmarks.* Surry Hills: Reader's Digest.

1985 (photographs by G. Matthews) *Rivers and Lakes in New Zealand.* Christchurch: Whitcoulls.

1987 'Lance William McCaskill: a tribute', *Proceedings of the 1987 Hill and High Country Seminar.* Lincoln: Tussock Grasslands and Mountain Lands Institute, Lincoln College, 237–56.

1990 'Landmarks revisited', transcript of an address to the Waikato Branch of the New Zealand Geographical Society on the occasion of the 21st AGM of the Branch, 17 November 1988. Hamilton: Department of Geography, University of Waikato/Waikato Branch, New Zealand Geographical Society.

2007 'Kenneth B. Cumberland: a memoir', *New Zealand Geographer* 63, 62–8.

2011 *Milestones and Landmarks.* Awhitu: Kenneth B. Cumberland.

2. OTHER REFERENCES CITED

Allan, R. S., Eccles, J. C., Forder, H. G., Packer, J., Parton, H. N., Popper, K. R. (1945), *Research and the University. A Statement by a Group of Teachers in the University of New Zealand.* Christchurch: Caxton Press.

Anderson, A. G. ed. (1980), *The Land Our Future. Essays on Land Use and Conservation in New Zealand in Honour of Kenneth Cumberland.* Auckland: Longman Paul.

Anderson, A. G., Kearns, R. A. and Hosking, P. L. (1996), *Fifty Years of Geography at The University of Auckland*, Occasional Publication No. 32. Auckland: Department of Geography, University of Auckland.

Bedford, R. (2011), 'Reflections on a region: New Zealand geographers and the study of Pacific peoples with special reference to the period 1945–1970', *New Zealand Geographer* 67, 126–38.

Buchanan, K. (1954), *Geography and Human Affairs: Inaugural Address.* Wellington: Victoria University College.

Buchanan, K. (1961), 'West wind, east wind', *New Zealand Geographer* 17, 189–96.

Buchanan, R. O. (1946), 'Accelerated erosion'. Review of *Soil Erosion in New Zealand: A Geographic Reconnaissance*, *New Zealand Geographer* 2, 272–4.

Bush, G. W. A. (1971), *Decently and in Order: The Government of the City of Auckland 1840–1971. The Centennial History of the Auckland City Council*. Auckland: Collins.

Clark, A. H. (1956), Review of *Southwest Pacific: A Geography of Australia, New Zealand and Their Pacific Island Neighbourhoods, Geographical Review* 46, 294–5.

Clark, A. H. (1960), Review of *New Zealand: A Regional View, Geographical Review* 50, 292–3.

Darby, H. C. (ed.) (1936), *An Historical Geography of England before A.D. 1800*. London: Cambridge University Press.

Downs, J. (1955), 'Impressions of the first New Zealand Geography Conference', *New Zealand Post-Primary Teachers' Association Journal* 1(7), 11.

Franklin, H. (1955), 'A cultural approach to geography', *Proceedings of the First Geography Conference*. Auckland: The New Zealand Geographical Society, A1, 1–9.

Gardner, W. J, Beardsley, E. T. and Carter, T. E. (1973), *A History of the University of Canterbury, 1873–1973*. Christchurch: University of Canterbury.

Hartshorne, R. (1939), *The Nature of Geography. A Critical Survey of Current Thought in the Light of the Past*. Lancaster PA: Association of American Geographers.

Jacks, G. V. (1945), 'Soil erosion in New Zealand', *Nature* 3972, 703.

Jobberns, G. (1945), 'Geography and national development', *New Zealand Geographer* 1, 5–18.

Lister, R. G. (1978), 'Retirement of Kenneth B. Cumberland: an appreciation', with Marshall, B. 'List of publications: by year of publication', *New Zealand Geographer* 34, 105–8.

Macaulay, J. U. (ed.) (1987), *Geographical Windows. Chronicle of the Geography Department, University of Canterbury, 1937–1987*. Christchurch: Department of Geography, University of Canterbury.

Mather, A. S. (1982), 'The changing perception of soil erosion in New Zealand', *The Geographical Journal* 148, 207–18.

Mayhew, R. J. (2011), 'Geography's genealogies', in J. A. Agnew and D. N. Livingstone (eds), *The SAGE Handbook of Geographical Knowledge*. London: Sage, 21–38.

McCaskill, L. W. (1973), *Hold This Land. A History of Soil Conservation in New Zealand*. Wellington: A.H. and A.W. Reed.

McCaskill, M. (1983), Review of *Landmarks, Australian Geographer*, 15, 348.

McCulloch, G. (1992), 'Defining and defending subjects: the case of the New Zealand Geographical Society', in G. McCulloch (ed.), *The School Curriculum in New Zealand: History, Theory, Policy and Practice*. Palmerston North: The Dunmore Press, 165–84.

McNaughton, A. H. (1955), 'Reflections on a geography conference', *National Education* 37, 298–9.

McSaveney, M. J. and Whitehouse, I. (1989), 'Anthropic erosion of mountain land in Canterbury', *New Zealand Journal of Ecology* 12 (Supplement), 151–63.

McWethy, D., Whitlock, C., Wilmshurst, J. M., McGlone, M. S., Fromont, M., Li, X., Dieffenbacher-Krall, A., Hobbs, W. O., Fritz, S. C. and Cook, E. R. (2010), 'Rapid landscape transformation in South Island, New Zealand, following initial Polynesian settlement', *Proceedings of the National Academy of Sciences of the United States of America* 107(50), 21343–8.

Moran, W. (2000), 'Classics in human geography revisited. Exceptionalism in the antipodes', *Progress in Human Geography* 24(3), 429–38.

O'Connor, K. F. (1980), 'The use of mountains: a review of the New Zealand experience', in A. G. Anderson (ed.), *The Land Our Future: Essays on Land Use and Conservation in New Zealand in Honour of Kenneth Cumberland*. Auckland: Longman Paul, 193–222.

Pawson, E. (2007), 'Landmarks: a personal view of the story of New Zealand', *Environmental History* 12, 350–1.

Pawson, E. (2011), 'Creating public spaces for geography in New Zealand: towards an assessment of the contributions of Kenneth Cumberland', *New Zealand Geographer* 67, 102–15.

Peden, R. (2011), *Making Sheep Country. Mt Peel Station and the Transformation of the Tussock Grasslands*. Auckland: Auckland University Press.

Perry, P. J. (1969), 'Twenty-five years of New Zealand historical geography', *New Zealand Geographer* 25, 93–105.

Popper, K. (1976), *Unended Quest. An Intellectual Biography*. Glasgow: Fontana/Collins.

Ray, P. (1981), 'The evolution of a frontman. Pauline Ray talks to Kenneth Cumberland, geographer, farmer and, at the age of 68, television presenter for *Landmarks*', *The Listener*, 22 August, 15–16.

Robinson, K. W. (1960), Review of *New Zealand: A Regional View*, *Australian Geographer* 8, 45–6.

Roche, M. (1997), '"The land we have we must hold": soil erosion and soil conservation in late nineteenth- and twentieth-century New Zealand', *Journal of Historical Geography* 23, 447–58.

Roche, M. (ed.) (2010), *A Geographer By Declaration. Selected Writings by George Jobberns*. Christchurch: Canterbury University Press.

Roche, M. (2011), 'George Jobberns and the writing of New Zealand geography: textbooks from the 1920s to the 1940s', *New Zealand Geographer* 67, 90–101.

Sauer, C. O. (1938), 'Theme of plant and animal destruction in economic history', *Journal of Farm Economics* 20, 765–75.

Sauer, C. O. (1941), 'Foreword to historical geography', *Annals Association of American Geographers* 31, 1–24.

Sauer, C. O. (1963), 'The morphology of landscape', in J. Leighly (ed.), *Land and Life. A Selection from the Writings of Carl Ortwin Sauer*. Berkeley: University of California Press, 315–50.

Soons, J. M. (2010), 'Jobberns, George – biography' [cited 8 December 2010]. Available at: www.TeAra.govt.nz/en/biographies/4j7/1

Stapledon, R. G. (1935), *The Land Now and Tomorrow*. London: Faber and Faber.

Stoddart, D. R. (1989), 'Epilogue: homage to Richard Hartshorne', in J. N. Entrikin and S. D. Brunn (eds), *Reflections on Richard Hartshorne's The Nature of Geography*. Washington, DC: Association of American Geographers, 163–6.

Symons, L. S. ('L.S.S.') (1960), Review of *New Zealand: A Regional View*, *Geography* 45, 145–6.

Van Dersal, W. R. (1947), 'Classics in the soil conservation field', *Journal of Water and Soil Conservation*, 2, 109–12.

Wilson, A. C. and Henry, M. G. (2011), 'Evolving disciplinary imaginaries: mapping the *New Zealand Geographer*, 1945–1969', *New Zealand Geographer* 67, 116–25.

Wilson, R. G. (1959), 'Mid-twentieth century New Zealand: geographers' viewpoint'. Review of *New Zealand: A Regional View*, *New Zealand Geographer* 15, 206–7.

Winder, G. and Lewis, N. (2010), 'Performing a new regional geography', *New Zealand Geographer* 66, 97–104.

Wright, J. K. (1952), *Geography in the Making. The American Geographical Society 1851–1951*. New York: American Geographical Society.

Chronology

1913	Born 1 October in Bradford
1932	Enters University College Nottingham
1936	Junior Lecturer in Geography, University College London
1938	Lecturer in Geography, Canterbury University College, Christchurch, New Zealand
1946	Senior Lecturer-in-Charge in Geography, Auckland University College
1949–78	Professorship in Geography, Auckland University College (from 1961, the University of Auckland)
2011	Dies 17 April, aged 97, in Auckland

Index

1. GENERAL INDEX

The index is divided into two parts:
1. A general index, including personal names, organizations, conferences, societies, and geographical concepts, theories and research.
2. A cumulative list of biobibliographies which includes all the geographers listed in volumes 1–30.

2. CUMULATIVE LIST OF BIOBIBLIOGRAPHIES